Applied Acoustics

G. Porges

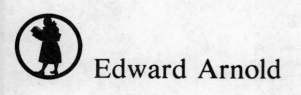

Edward Arnold

© G. Porges 1977

First published 1977
by Edward Arnold (Publishers) Limited
25 Hill Street, London W1X 8 LL

British Library Cataloguing in Publication Data

Porges, G
    Applied acoustics.
    1. Acoustical engineering
    I. Title
    620.2   TA365

    ISBN 0–7131–2658–2
    ISBN 0–7131–2659–0 Pbk

Text set in 10/11 pt IBM Press Roman,
printed by photolithography, and bound
in Great Britain at The Pitman Press, Bath

Preface

Noise appears in many practical problems, and a number of formulae and theories have been found useful in coping with these problems. I have set myself the task of developing the more important and useful of them from first principles in the simplest possible way.

Acoustics is a specialist subject. This book provides an introductory text which assumes no previous knowledge of sound or other waves. While it makes use of maths and physics at about first year University standard, the emphasis is on practical results and applications.

The book will be of use to students on the relatively new courses in environmental subjects, as well as to practising engineers and physicists who have gained their basic qualifications in other branches of their discipline. To meet their needs enough of the physical theory of sound has been given to show that the results quoted are valid, but I have not presumed on the reader's interest in pure mathematics.

Since my object is the practical application of acoustic theory, I have tried not to obscure the physical significance with mathematical symbolism. For example, I have deliberately worked in terms of frequency f rather than angular frequency $\omega = 2\pi f$ which has no obvious significance for sound waves.

I would have liked to keep the mathematics to the level of simple trigonometry and calculus, and have done so as far as possible but there are some sections where I have felt justified in using the more concise notation of complex algebra. I have included a short section on cylindrical waves, in spite of the fact that this has made it necessary to refer to Bessel functions, which are beyond the level of the rest of the mathematics.

In dealing with spherical waves, I do not quote the form of the wave in spherical co-ordinates without proof, as is so often done, but derive it afresh. I also explain why the simplified reasoning which relates intensity and energy density of plane waves does not apply to other waves.

The same equations appear in acoustics, vibrations, electrical engineering, control theory and other fields, and it has been possible to develop analogies between different subjects. These analogies make it possible to manipulate the equations without regard to their significance. This is a poor use of such similarities. A better use requires the author to assume that his readers have more knowledge of one subject, usually electrical engineering, than of others. I am unwilling to make such an assumption, and so have avoided purely mathematical analogies.

Numerical data for acoustic calculations are not always easy to come by. Some have therefore been included in the appendix, so that the book may serve not only as an introduction to the subject, but as a practical manual.

I should like to express my thanks to Mrs A. C. Porges for her help in the preparation of the manuscript.

London
1977

G. Porges

Contents

77

List of symbols

a	arbitrary constant, equivalent absorption
b	arbitrary constant
c	velocity of sound
d	distance, deflection
f	frequency
k	wave number $= 2\pi/\lambda$
l	length
m	mass, mass per unit area
p	pressure
r	radial co-ordinate
s	displacement, spring rate
t	time
u	particle velocity
v	velocity
w	mass, power per unit solid angle
x	co-ordinate, deflection
y	co-ordinate
z	co-ordinate

A	area, arbitrary angle, distance, constant
B	arbitrary angle, distance, constant
E	modulus of elasticity, energy
F	force
I	intensity
IL	intensity level
P	pressure, phons, perimeter length, function defined in text
PNL	perceived noise level
Q	directivity factor, function defined in text
R	gas constant, function of acoustic resistance, room constant, damping coefficient
S	sones, area
SPL	sound pressure level
SRI	sound reduction index
SWL	sound power level
T	periodic time, reverberation time, temperature, force transmission ratio, function of t
V	volume
W	power
X	acoustic reactance

α	coefficient, real part of complex number, absorption coefficient
β	constant, imaginary part of complex number
γ	ratio of specific heats
θ	arbitrary angle, phase angle, co-ordinate, temperature
λ	wavelength
ρ	density
σ	surface density
τ	transmission coefficient
ϕ	phase angle
ψ	co-ordinate
$\mathscr{E}$	energy per unit volume

1 The nature and velocity of sound

1.1 Sounds and waves

The concepts of power and energy are essential in understanding the more practical applications of acoustics. To say that sound is a form of energy may not be a conventional dictionary definition of an ordinary English word, but it conveys quickly and simply one of the essential features of sound. The other essential feature is that sound is always associated with vibration of material particles. Energy is transferred from one vibrating particle to the next, and acoustic energy travels through the medium as a wave.

The local displacement of the particles of the medium produces a local **compression** followed rapidly by a local **rarefaction** followed equally rapidly by another local compression, and so on. This series of compressions and rarefactions spreads through the medium. The study of sound is concerned with the study of this wave motion. It must be emphasized and clearly understood that the movement of the wave is different and quite distinct from the movement of the individual particles. This is illustrated in fig. 1.1, and it can be seen very clearly in the motion of the sea at the seaside. There is a definite wave motion consisting of waves moving towards the shore. Anyone swimming through the waves is in no doubt whether he is on the crest or in the trough of the wave. Nor is there any doubt that the wave is moving forward. But if you place a cork or a piece of seaweed in the water and watch it, you will notice that although it moves up and down, it will not move forward with the wave (unless it is very close to the shore, where the depth of water becomes less than the height of the wave). As the wave crest moves on, the cork drops into the following trough but remains at the same distance from the shore as when it was on the crest. As it rises to the next crest, it still does not move horizontally. This is an example of a **transverse wave** in which the local particle displacement takes place at right angles to the direction in which the wave is moving.

Sound generally travels in **longitudinal waves** in which the small local particle displacement takes place in the same direction as the wave movement, as shown in fig. 1.2. This type of wave motion can be demonstrated with a fairly long light spring. If one end of the spring is held in one hand and two or three coils at the other end are displaced, a ripple will be seen to travel the length of the spring. The ripple is formed by adjacent coils of the spring coming closer together and then moving apart again. Each coil makes a small oscillation about its equilibrium position. It is the compression of a small length of spring that travels along the spring, while the coils themselves return to their normal position. If the spring is suitably supported it will also demonstrate the reflection of the wave from the ends. Sound waves are reflected in an analogous manner, as will be discussed later.

The mathematics of the two types of waves are identical. Both are accurately represented by the same equations and it will be found that in much of what follows it is unnecessary to distinguish between them. The wave of compression and rarefaction is transmitted through the medium because the material possesses elasticity as well as inertia or mass. Sound is a material vibration whose propagation

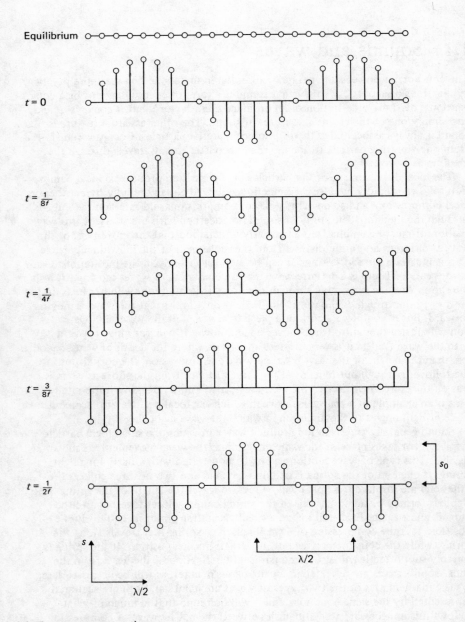

1.1 Transverse wave motion

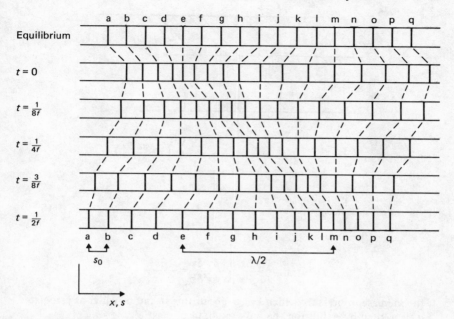

1.2 Longitudinal wave motion

depends on these physical properties of matter. Unlike light it cannot be transmitted through a vacuum which has neither inertia nor elasticity. We can derive an expression for the velocity of sound through matter from a consideration of these properties.

1.2 The wave equation

First we will consider particle displacement, acoustic pressure, and the relationship between them. This relationship will then be substituted in the Newtonian laws of motion. Since displacement and pressure vary with both distance and time, partial differentials must be used.

In fig. 1.3, consider a longitudinal compression wave travelling in the x direction through an infinitesimal element having dimensions $\delta x\,\delta y\,\delta z$. Suppose the centre of gravity is displaced in the positive x direction by a distance s. Then the boundaries will be displaced by $\left(s - \dfrac{\partial s}{\partial x}\dfrac{\delta s}{2}\right)$ and $\left(s + \dfrac{\partial s}{\partial x}\dfrac{\delta x}{2}\right)$ respectively. The volume therefore decreases by

$$\left[\left(s - \frac{\partial s}{\partial x}\frac{\delta x}{2}\right) - \left(s + \frac{\partial s}{\partial x}\frac{\delta x}{2}\right)\right]\delta y\,\delta z = -\frac{\partial s}{\partial x}\delta y\,\delta z\,\delta x$$

The volumetric strain is defined as the ratio of the decrease in volume to the original volume, so we have

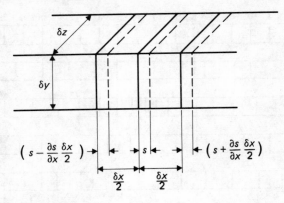

$$\left(s - \frac{\partial s}{\partial x} \frac{\delta x}{2} \right) \rightarrow \quad \leftarrow \quad \rightarrow s \leftarrow \quad \rightarrow \quad \leftarrow \left(s + \frac{\partial s}{\partial x} \frac{\delta x}{2} \right)$$

$$\frac{\delta x}{2} \qquad \frac{\delta x}{2}$$

1.3 Plane wave displacement

$$\text{strain} = -\frac{\partial s}{\partial x} \frac{\delta x\, \delta y\, \delta z}{\delta x\, \delta y\, \delta z} = -\frac{\partial s}{\partial x}$$

If the increase in pressure which brings about this strain, or which is associated with it, is p, then by definition the bulk modulus of elasticity is

$$E = \frac{\text{stress}}{\text{strain}} = \frac{-p}{\partial s/\partial x}$$

$$\therefore \quad p = -E\, \partial s/\partial x$$

Now let us consider the force on, and the acceleration of, the element. The force in the x direction is due to the difference in pressure on the faces at $s - \delta x/2$ and at $s + \delta x/2$. This difference in pressure is $(\partial p/\partial x)\delta x$ and it acts on an area $\delta y\, \delta z$. Therefore the force is $(\partial p/\partial x)\, \delta x\, \delta y\, \delta z$ acting in the negative x direction.

The displacement of the element is s, so its acceleration is $\partial^2 s/\partial t^2$. Its mass is $\rho\, \delta x\, \delta y\, \delta z$ where ρ is the density. Therefore, from the laws of motion,

$$-\frac{\partial p}{\partial x} \delta x\, \delta y\, \delta z = \rho\, \frac{\partial^2 s}{\partial t^2}\, \delta x\, \delta y\, \delta z$$

Substituting

$$\frac{\partial p}{\partial x} = -E\, \frac{\partial^2 s}{\partial x^2}$$

we have

$$-E\, \frac{\partial^2 s}{\partial x^2} = -\rho\, \frac{\partial^2 s}{\partial t^2}$$

$$\therefore \quad \frac{\partial^2 s}{\partial t^2} = \frac{E}{\rho}\, \frac{\partial^2 s}{\partial x^2} \tag{1.1}$$

This is the well known differential **wave equation** of a periodic fluctuation in s which is propagated in the x direction at a velocity

$$c = \sqrt{(E/\rho)}$$

Let us consider a solution of the form

$$s = s_0 \cos 2\pi f(t - x/c) \qquad (1.2)$$

where f is any frequency.

Then

$$\frac{\partial s}{\partial t} = -2\pi f s_0 \sin 2\pi f(t - x/c)$$

and

$$\frac{\partial^2 s}{\partial t^2} = -4\pi^2 f^2 s_0 \cos 2\pi f(t - x/c) \qquad (1.3)$$

Also

$$\frac{\partial s}{\partial x} = \frac{2\pi f}{c} s_0 \sin 2\pi f(t - x/c)$$

and

$$\frac{\partial^2 s}{\partial x^2} = -\frac{4\pi^2 f^2}{c^2} s_0 \cos 2\pi f(t - x/c) \qquad (1.4)$$

By substituting eqns 1.3 and 1.4 in eqn 1.1, we confirm that eqn 1.2 from which they are derived is a solution of eqn 1.1.

We have obtained a wave equation for the displacement of the element. A corresponding equation can be obtained for the local acoustic pressure. We go back to the equation of motion

$$\frac{-\partial p}{\partial x} = \rho \frac{\partial^2 s}{\partial t^2}$$

$$\therefore \quad \frac{\partial^2 p}{\partial x^2} = -\rho \frac{\partial^3 s}{\partial x \, \partial t^2}$$

As before, we use the stress/strain relationship

$$p = -E \frac{\partial s}{\partial x}$$

$$\therefore \quad \frac{\partial^2 p}{\partial t^2} = -E \frac{\partial^3 s}{\partial x \, \partial t^2}$$

Comparing the last two results gives

$$\frac{\partial^2 p}{\partial t^2} = \frac{E}{\rho} \frac{\partial^2 p}{\partial x^2} \qquad (1.5)$$

which is the same wave relation for p with the same velocity, as we have already derived for s.

We thus see that the **velocity** of propagation of the sound wave is related to the properties of the material by the expression

$$c = \sqrt{(E/\rho)} \qquad (1.6)$$

Some discussion of the wave equation in the form $s = s_0 \cos 2\pi f(t - x/c)$ will not be out of place to show that it does represent a travelling wave. The equation is represented in fig. 1.4. We can consider what happens at any one point by putting $x = 0$. The equation then reduces to $s = s_0 \cos 2\pi ft$, at any fixed point the disturbance varies sinusoidally with time. If we put $t = 0$ then $s = s_0 \cos 2\pi fx/c$, at any fixed time the disturbance is spread out sinusoidally through space.

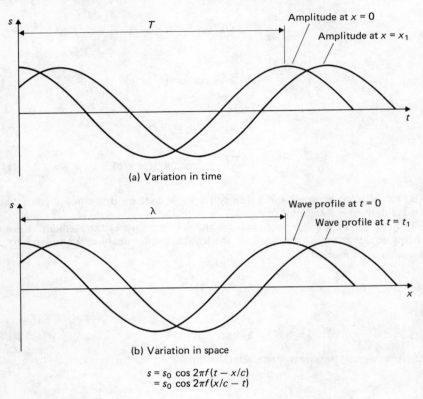

(a) Variation in time

(b) Variation in space

$$s = s_0 \cos 2\pi f(t - x/c)$$
$$= s_0 \cos 2\pi f(x/c - t)$$

1.4 Representation of a travelling wave

The maximum value of s is s_0, and occurs at $f(t - x/c) = n$ where n is 0 or any integer. Taking the simplest case of $n = 0$, we have $t = x/c$ or $x = ct$. This equation tells us that the value of x at which the maximum occurs increases steadily with time t, i.e. the crest moves forward at a speed c. If we take n as an integer, we get $x = ct +$ constant which still represents a point of constant value s moving in the x direction at a speed c. The same result can be obtained for the minimum value of s or for any intermediate value. The only difference is the value of the constant in $x = ct +$ constant.

The **periodic time** of the fluctuation at $x = 0$ is found by putting $s_0 \cos 2\pi f(t + T)$ $= s_0 \cos 2\pi ft$ from which $2\pi fT = 2\pi$ and $T = 1/f$. The **wavelength** at $t = 0$ is found by putting $s_0 \cos 2\pi fx/c = s_0 \cos 2\pi f(x + \lambda)/c$ from which $2\pi f\lambda/c = 2\pi$ and $c = \lambda f$.

1.3   Wavelength and frequency

What we have just expressed mathematically can also be described physically by thinking again about the waves at the seaside. We can describe the wave motion which we see by the distance between the successive crests, or between the successive troughs. It is this distance or length which is called the wavelength.

Let us consider a wave of wavelength λ metres which is moving forward at a speed c metres per second and let us measure the time in seconds from the instant when one crest passes a fixed point in space. Since the point is fixed we can take its position as being given by $x = 0$. Suppose T seconds pass before the next crest passes the same point. Then, by definition, the wave has travelled λ metres in T seconds. Therefore its velocity is

$$c = \lambda/T \, \text{ms}^{-1}$$

But the time T is just the time required for the local displacement to come back to its original value, i.e.

$$s_0 \cos 2\pi f T = s_0 \cos 0$$

$$\therefore \ 2\pi f T = 0 \text{ or } 2\pi$$

$$\therefore \quad T = 1/f$$

This provides a relationship between velocity, wavelength, and frequency:

$$c = \lambda/T$$

$$\therefore \ c = \lambda f$$

We have already discussed how the velocity c depends on the properties of the material. What happens when sound passes from one material to another? If c changes, either f or λ must also change, or the relationship between the three would cease to hold good. Imagine the wave as originating from the prongs of a tuning fork. Then we can see that the frequency of the ups and downs will remain the same as the wave passes from one medium to another. For consider what would happen if it were not so. At some time a compression or crest on one side of the boundary would coincide with a rarefaction or trough on the other side. This is so inconsistent with continuous motion that we must regard it as impossible. On the other hand there is nothing to prevent the crests bunching up together and so reducing the wavelength.

The same reasoning can be expressed in more mathematical language by saying that the condition of continuity at the boundary requires that the displacements on either side of it are equal at all times, i.e. at the point $x = 0$

$$s_0 \cos 2\pi f_1 t = s_0 \cos 2\pi f_2 t$$

This is only possible for all values of t if $f_1 = f_2$. So for any one sound, frequency f is constant while both the wavelength and speed depend on the nature of the material through which the sound is travelling.

Frequency and amplitude are the two factors which principally characterize a wave motion. As a very rough approximation it may be said that the ear detects differences in frequency and amplitude as differences of pitch and loudness respectively. The correlation is far from exact, and one of the difficulties in practical acoustics is trying to get a correlation which bears some resemblance to the behaviour of the human ear.

1.4   Velocity of sound in air

Since the commonest path for sound waves is through the air, it is worth applying the general expression for the velocity of sound to the particular case of air. Air can be treated as a perfect gas, for which $PV = wRT$ where R is the gas content for one kilogram-mol. Here P is the absolute pressure, whereas p was the acoustic over pressure or the pressure variation taking place round the steady value P. We could write $p = dP$, but this would make the notation unnecessarily cumbersome.

The sound travels so rapidly that in all ordinary cases there is no time for any heat transfer to take place, and so the changes are considered adiabatic. For an adiabatic change in a perfect gas

$$PV^\gamma = \text{constant}$$

Differentiating with respect to V

$$V^\gamma \frac{dP}{dV} + \gamma PV^{\gamma-1} = 0$$

$$\therefore \qquad V^\gamma \frac{dP}{dV} = -\gamma PV^{\gamma-1}$$

Dividing by $V^{\gamma-1}$,

$$\frac{V}{dV} dP = -\gamma P$$

dV is positive change in volume, so the decrease in volume is $-dV$, and the volumetric strain is $-dV/V$. The stress associated with the strain is dP, so for the adiabatic bulk modulus of elasticity we have $E = -V \, dP/dV = \gamma P$

The density ρ of a gas is $\dfrac{m}{V} = \dfrac{P}{RT}$

so

$$c = \sqrt{(E/\rho)}$$
$$= \sqrt{(\gamma PRT/P)}$$
$$= \sqrt{(\gamma RT)}$$

What we have shown is that the velocity of sound in air is independent of pressure, but that it does vary with the temperature, and is in fact proportional to the square root of the absolute temperature. This is an important result when we consider how sound travels through the atmosphere.

1.5   Phase difference

For the sake of completeness it should be added that otherwise identical waves can be out of step with each other, or differ in phase. Eqn 1.2 should be modified to read $s = s_0 \cos{[2\pi f(t - x/c) + \phi]}$ where ϕ is a phase angle. Fortunately it is found that in sound waves the physical effect of phase is frequently unimportant and may often be ignored.

2 Sound pressure, power and intensity

2.1 Octave bands

So far we have been considering a single sound of one frequency only. This would be such a pure tone that it would be unpleasant to listen to. Every musical instrument produces a series of overtones or harmonics in addition to the pure note actually played. The frequencies of the harmonics are multiples or fractions of the fundamental frequency, and one of the major differences between different types of musical instrument is the range of harmonics accompanying each fundamental note.

But it is not only music that can consist of several different frequencies sounding

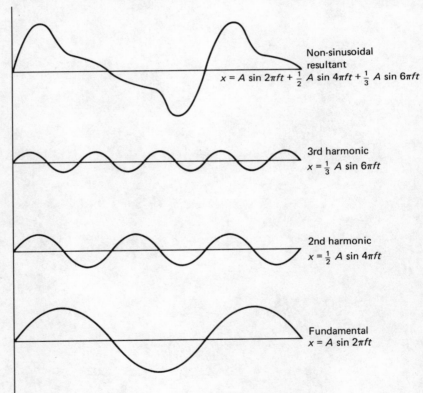

Non-sinusoidal resultant
$$x = A \sin 2\pi ft + \tfrac{1}{2} A \sin 4\pi ft + \tfrac{1}{3} A \sin 6\pi ft$$

3rd harmonic
$$x = \tfrac{1}{3} A \sin 6\pi ft$$

2nd harmonic
$$x = \tfrac{1}{2} A \sin 4\pi ft$$

Fundamental
$$x = A \sin 2\pi ft$$

2.1 Example of a Fourier series

together. Any continuing vibration can be expressed as a Fourier series of pure sine waves of differing frequency, amplitude and phase. Fig. 2.1 illustrates one example of a non-sinusoidal wave which is made up of three different sine waves. The frequencies audible to the human ear range from about 20 hertz to about 20 000 hertz, one hertz being one cycle per second. To be heard, even a very deep or low frequency sound must last for several cycles. During this time it may be heard together with any number of other sounds, all of different frequency and amplitude. How pleasant or unpleasant the resulting sound is depends on the exact combination of pure tones.

It has already been pointed out that the ear reacts differently to different frequencies. It is therefore important that we should always take into account the actual frequency of a sound, or the frequencies of the component sounds. Now we are interested in a continuous spectrum of frequencies ranging from 20 to 20 000 Hz and to avoid having to do a vast number of calculations for a vast number of frequencies, it is normal to divide the range into octave bands. The top frequency of each band is twice the bottom frequency of that band, and the top frequency of one band is of course the bottom frequency of the next. The precise limits of the octave bands are quite arbitrary, and not everyone has always used the same figures. A set of values which was quite common is:

Band 1	20 to	75 Hz	mid-frequency	53 Hz
2	75 to	150		106
3	150 to	300		212
4	300 to	600		425
5	600 to	1200		850
6	1200 to	2400		1700
7	2400 to	4800		3400
8	4800 to.	10000		6800

It will be seen that, for example, in band 4, $600 = \sqrt{2} \times 425$ and $425 = \sqrt{2} \times 300$. Bands 1 and 8 have been extended slightly at the lower and upper ends respectively to increase the range of frequencies covered. Although the ends of the bands are round numbers, this series gives awkward numbers for the mid-frequencies, which are used far more often in calculation and measurement than the band-widths. If the mid-frequencies at the lower end are rounded off, the series becomes: 50, 100, 200, 400, 800, 1600, 3200 and 6400 Hz.

The commonest series of mid-frequencies now used is: 63, 125, 250, 500, 1000, 2000, 4000, 8000 Hz. This is the preferred series of B.S. 3593.

To give some reality to this list of numbers it is as well to remember that middle C has a frequency of about 256 Hz. The notes on a piano keyboard follow the same law of geometrical progression as has been adopted for defining octave bands, so that the C above middle C has a frequency of about 512 Hz, and that below about 128 Hz. Thus the numbers that one reads on a frequency meter do bear some relation to the ordinary piano keyboard.

For most acoustic problems it is enough to specify a sound by average values for each octave band. For more precise work each octave is divided into three and we then work in third-octave bands.

2.2   Sound pressure

We have described how sound travels as a wave of compression and rarefaction with an associated wave of pressure variation. In most practical problems it is the pressure variation that is of greater importance and of greater interest. The acoustic pressure at any point is the difference between the actual pressure at that point in the presence of the sound and the pressure that would exist at that point, under identical conditions, in the absence of any sound. This acoustic over pressure at any point varies sinusoidally with time, exactly as an alternating electric current, and, exactly as in electrical measurements, it is convenient to use the root mean square, or r.m.s., value. This is defined as the square root of the average of the squares of the instantaneous pressures. In mathematical notation,

$$p_{r.m.s.}^2 = \frac{1}{T} \int_0^T p^2 \, dt$$

For a sine wave of the sort derived in Chapter 1,

$$p = p_0 \cos 2\pi (ft - x/\lambda) \tag{2.1}$$

where p_0 is the maximum value of p.

Then

$$p_{r.m.s.}^2 = \frac{1}{T} \int_0^T p_0^2 \cos^2 2\pi (ft - x/\lambda) \, dt$$

$$= \frac{p_0^2}{2T} \int_0^T [1 + \cos 4\pi (ft - x/\lambda)] \, dt$$

$$= \frac{p_0^2}{2T} \int_0^T dt + \frac{p_0^2}{2T} \int_0^T \cos 4\pi (ft - x/\lambda) \, dt$$

The last term represents the time average of a cosine, and this is zero. The first term integrates to $1/2 \, p_0^2$ so we have

$$p_{r.m.s.}^2 = p_0^2/2 \tag{2.2}$$

or

$$p_{r.m.s.} = p_0/\sqrt{2}$$

Fig. 2.2 illustrates these relations graphically.

If two sounds of different frequency act together, then the total r.m.s. sound pressure p_t is found as follows:

$$p_t = p_1 + p_2 = p_{10} \cos 2\pi f_1 (t - x/c) + p_{20} \cos 2\pi f_2 (t - x/c)$$

therefore

$$p_t^2 = p_{10}^2 \cos^2 2\pi f_1(t - x/c) + p_{20}^2 \cos^2 2\pi f_2(t - x/c)$$
$$+ 2p_{10}p_{20} \cos 2\pi f_1(t - x/c) \cos 2\pi f_2(t - x/c)$$
$$= \frac{p_{10}^2}{2} + \frac{p_{10}^2}{2} \cos 4\pi f_1(t - x/c) + \frac{p_{20}^2}{2} + \frac{p_{20}^2}{2} \cos 4\pi f_2(t - x/c)$$
$$+ p_{10}p_{20} \cos 2\pi (f_1 + f_2)(t - x/c) + p_{10}p_{20} \cos 2\pi (f_1 - f_2)(t - x/c)$$

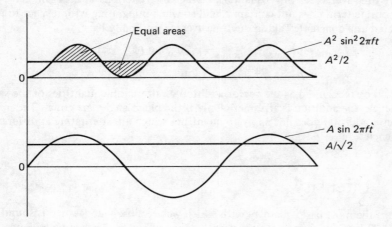

2.2 R.M.S. value of a sine wave

To find the r.m.s. value of the total pressure, we again take the time average, and in this process the cosine terms disappear, so that we may write

$$\frac{1}{T} \int_0^T p_t^2 \, dt = \frac{1}{2T} \int_0^T p_{10}^2 \, dt + \frac{1}{2T} \int_0^T p_{20}^2 \, dt$$

Using the results already obtained, this reduces to

$$p_{t,\text{r.m.s.}}^2 = p_{1,\text{r.m.s.}}^2 + p_{2,\text{r.m.s.}}^2 \qquad (2.3)$$

If a sound covers a continuous band of frequencies, then by extending the above analysis we will arrive at the result

$$p_{t,\text{ r.m.s.}}^2 = \int p_{\text{r.m.s.}}^2 \, df$$

This illustrates a matter of fundamental importance, namely that r.m.s. sound pressures of different frequencies cannot be added linearly or arithmetically. Suppose a noise is due to two sounds which are of equal pressure but different frequency. Then the total r.m.s. sound pressure is not twice, but $\sqrt{2}$ times, that due to either sound alone.

It will be found that if a phase angle had been included it would not have affected the previous result. If, however, the two sounds are of equal frequency, we have one of the few cases where phase angle does matter. The analysis of this case is as follows:

$$p_t^2 = p_{10}^2 \cos^2 2\pi \left[(ft - x/\lambda) + \phi_1\right] + p_{20}^2 \cos^2 2\pi \left[(ft - x/\lambda) + \phi_2\right]$$

$$+ 2p_{10}p_{20} \cos 2\pi \left[(ft - x/\lambda) + \phi_1\right] \cos 2\pi \left[(ft - x/\lambda) + \phi_2\right]$$

$$= p_{10}^2/2 + (p_{10}^2/2) \cos 4\pi \left[(ft - x/\lambda) + \phi_1\right] + p_{20}^2/2 + (p_{20}^2/2) \cos 4\pi \left[(ft - x/\lambda) + \phi_2\right]$$

$$+ p_{10}p_{20} \cos 2\pi \left[(2ft - 2x/c) + \phi_1 + \phi_2\right] + p_{10}p_{20} \cos 2\pi (\phi_1 - \phi_2)$$

The time average of the cosine terms which are functions of t is again zero. The other cosine term does not contain t, and so when integrating with respect to t it is treated as a constant. Taking the time average we finally have

$$p_{t,\,\text{r.m.s.}}^2 = p_{1,\text{r.m.s.}}^2 + p_{2,\text{r.m.s.}}^2 + p_{10}p_{20} \cos(\phi_1 - \phi_2)$$

$$= p_{1,\text{r.m.s.}}^2 + p_{2,\text{r.m.s.}}^2 + 2p_{1,\text{r.m.s.}}\ p_{2,\text{r.m.s.}} \cos(\phi_1 - \phi_2) \tag{2.4}$$

This will be recognised as the vector addition of alternating quantities of the same frequency, the addition is arithmetic only if the phase angles are equal. The method of vector addition does not apply to quantities which are alternating at different frequencies.

2.3 Intensity

Perhaps the most basic quantity with which we are closely concerned is sound power. This is associated with the actual source of sound. The source radiates power which is transmitted in the form of sound. The sound power of a source is the total power coming from it. It is the rate at which energy in the form of sound leaves the source.

Now consider a point some distance from the source, and a small area perpendicular to the line joining the point to the source. Then some of the power being generated by the source will be transmitted through the area, the exact amount depending not only on the sound power of the source but also on its directional properties, the distance of the area from the source, and the presence of sound absorbing or sound reflecting materials. If the power passing through an area A is W, then we define the intensity I as power per unit area, or

$$I = W/A$$

Although the intensity of a sound is an important quantity, it is difficult to measure, whereas sound pressure can be measured quite readily. There is, however, a relationship between the two. This is

$$I = p_{\text{r.m.s.}}^2/\rho c$$

where ρ is the density of the air or other medium and c is the velocity of sound in the medium.

This relationship between pressure and intensity can be derived by considering the sound wave as carrying or transmitting energy. We have described how each infinitesimal particle of the medium through which the sound is travelling vibrates. There is both potential and kinetic energy associated with the vibration.

The potential energy of an element of volume V is $-\int p\,dV$, the minus sign appearing because an increase of pressure is accompanied by a decrease of volume and an increase of energy. For a longitudinal compression wave travelling in the x-direction, the volume is

$$V = V_0 \left(1 + ds/dx\right)$$

where V_0 is the steady state volume, and s is the displacement in the x-direction, and ds/dx is the volumetric strain.
But it has been shown that for such a wave

$$ds/dx = -p/E = -p/c^2\rho$$
$$\therefore \ V = V_0 \left(1 - p/c^2\rho\right)$$

Differentiating gives

$$dV = -V_0 \, dp/c^2\rho$$

so that the potential energy is

$$E_p = -\int p \, dV = \int \frac{V_0}{c^2\rho} p \, dp$$
$$= \tfrac{1}{2} \frac{p^2}{c^2\rho} V_0$$

The constant of integration is zero because we take $E_p = 0$ when $p = 0$.
   The kinetic energy is simply $E_k = \tfrac{1}{2}\rho u^2 V_0$ where u is the velocity of displacement.
   The total energy E associated with the volume V_0 is the sum of E_p and E_k.
Before adding these we need a relationship between p and u.
   If we express the displacement s as

$$s = s_0 \cos 2\pi(ft - x/\lambda)$$

we have

$$p = -\rho c^2 \, ds/dx$$
$$= -\rho c^2 s_0 \frac{2\pi}{\lambda} \sin 2\pi(ft - x/\lambda)$$
$$= -\rho c \, 2\pi f s_0 \sin 2\pi(ft - x/\lambda)$$

The velocity is

$$u = ds/dt$$
$$= -2\pi f s_0 \sin 2\pi(ft - x/\lambda)$$

Comparing the last two expressions, we see that

$$p = \rho c u \tag{2.5}$$

(An alternative derivation of this equation is given in Appendix 1).

We can now add the two energy terms,

$$E = E_\mathrm{p} + E_\mathrm{k}$$

$$= \tfrac{1}{2} \frac{p^2}{c^2 \rho}\; V_0 + \tfrac{1}{2}\rho u^2 V_0$$

$$= \tfrac{1}{2}\,\rho V_0 \left(\frac{p^2}{c^2 \rho^2} + \frac{p^2}{c^2 p^2} \right)$$

$$= p^2 V_0 / \rho c^2$$

This is the energy of acoustic vibration of a small volume V_0.

We can define energy density as the energy per unit volume,

$$\mathscr{E} = E/V_0 = p^2/\rho c^2 \tag{2.6}$$

Now the intensity of the sound is the average rate at which energy flows through a unit area perpendicular to the direction of travel. Since the pressure and particle velocity are propagated as longitudinal waves, so is the energy. The energy which, at any instant, is contained in a column of unit cross sectional area and of length $c\,dt$ will pass through the end of the column in a time dt fig. 2.3. Therefore the energy passing in time dt is $\mathscr{E}c\,dt$. Therefore the average rate of flow of energy, the intensity, is $\mathscr{E}c$. Hence $I = \mathscr{E}c = p^2/\rho c$.

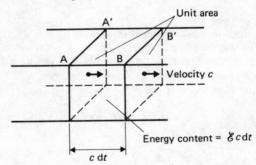

2.3 Derivation of intensity

We can readily show that this relationship holds for mean values as well as for instantaneous values.

$$I = \frac{p^2}{\rho c} = \frac{p_0^2}{\rho c} \cos^2 2\pi (ft - x/\lambda)$$

$$= \tfrac{1}{2}\frac{p_0^2}{\rho c} + \tfrac{1}{2}\frac{p_0^2}{\rho c} \cos 4\pi (ft - x/\lambda)$$

and

$$\frac{1}{T}\int_0^T I\,dt = \frac{1}{2T\rho c}\int_0^T p_0^2\,dt + \frac{1}{2T\rho c}\int_0^T p_0^2 \cos 4\pi (ft - x/\lambda)\,dt$$

or

$$I_{\text{mean}} = \rho_0^2/2\rho c = p_{\text{r.m.s.}}^2/\rho c$$

The intensity, although fluctuating, is always positive. This is clear from a consideration of the physical meaning and from the mathematics. Since I is proportional to p^2 which must always be positive the time average of the intensity will also be positive. This mean value of intensity is a useful quantity.

The pressure, on the other hand, fluctuates evenly about zero. The time average is therefore zero, and an attempt at using a simple mean value conveys no useful information. This is why, when dealing with pressure, we have to use the root mean square value rather than the ordinary mean value which we can use for intensity.

All this may be summarized by saying that a source generates sound energy and the rate at which it does so is the sound power radiated by it. As the power flows from the source, the amount passing through unit area in unit time is the intensity. The power travels as a pressure wave whose magnitude is the sound pressure. The following equation has been derived to relate the intensity to the sound pressure

$$I = p_{\text{r.m.s.}}^2/\rho c$$

2.4   Decibels

Just as the frequencies in which we are interested cover a large range, so do the actual values of sound pressure, intensity and power, and it is therefore convenient to measure them on a logarithmic scale. The scale chosen for this is the **decibel scale**.

The word decibel is in practice used very loosely, and many people using it do not know what it really means. It is important to understand quite clearly that the decibel is fundamentally a ratio of powers. It is in fact a useful scale for electrical quantities, in spite of many people's associations of the word decibel with sound.

If we wish to compare two powers w_1 and w_2 watts, the definition is that

$$10 \log (w_1/w_2) = \text{decibel level of } w_1 \text{ above } w_2,$$

where the logarithms are taken to base ten and the numerical value is given in decibels, dB.

It will be remembered that electrical power is proportional to the square of current i or to the square of voltage v. Hence we have

$$\text{level in dB} = 10 \log (i_1^2/i_2^2) = 10 \log (i_1/i_2)^2$$

$$= 20 \log (i_1/i_2)$$

and similarly

$$\text{level in dB} = 20 \log (V_1/V_2)$$

Note that a particular value of the decibel level does not specify the value of a current or a voltage unless we also specify or imply a reference current or voltage with which all values are to be compared.

The decibel scale is applied to sound measurements by means of the following definitions:

$$\text{sound power level SWL} = 10 \log (W/W_0) \text{ dB re } W_0$$

$$\text{intensity level IL} = 10 \log (I/I_0) \text{ dB re } I_0$$

$$\text{sound pressure level SPL} = 10 \log (p^2/p_0^2)$$

$$= 20 \log (p/p_0) \text{ dB re } p_0$$

In these definitions all quantities are r.m.s. values

The word level implies that the quantity is being measured as a ratio to a special reference magnitude, and the ratio is expressed in decibels. The terms re W_0, re I_0, and re p_0 simply mean that we are using W_0, I_0, and p_0 as standard reference values with respect to which all our measurements are going to be made. The usual values used in acoustics are:

$$W_0 = 10^{-12} \text{ W}$$

$$I_0 = 10^{-12} \text{ W m}^{-2} \quad = 10^{-16} \text{ W cm}^{-2}$$

$$p_0 = 0.00002 \text{ N m}^{-2} = 0.0002 \text{ dyn cm}^{-2}$$

$$= 20 \ \mu\text{Pa (micropascal)}$$

Unless specifically stated otherwise it may be assumed that these are the reference values being used, but care is needed because in American practice the standard of W_0 has sometimes been taken as 10^{-13} watts. It can now be seen that a measurement given in decibels is completely meaningless unless we are also told, either explicitly or by implication, what quantity is being measured and with respect to what reference level.

We can derive a numerical version of eqn 2.6 for the special, but very important, case of air at standard temperature and pressure (20°C, 1.01×10^5 pascal), when the density ρ is 1.21 kg m^{-3}, and the acoustic velocity is 343 m s^{-1}. This gives the product $\rho c = 415$ rayls or 415 kg m^{-2} s^{-1}. Hence

$$I = p^2/415 = (p/20.37)^2$$

$$\therefore \quad \frac{I}{10^{-12}} = \left(\frac{p}{20.37 \times 10^{-6}}\right)^2 = \left(\frac{p}{2.037 \times 10^{-5}}\right)^2$$

$$= \left(\frac{p}{2 \times 10^{-5}}\right)^2 \times \left(\frac{1}{1.019}\right)^2$$

$$\therefore \quad 10 \log \frac{I}{10^{-12}} = 20 \log \left(\frac{p}{2 \times 10^{-5}}\right) - 20 \log 1.019$$

$$\therefore \quad \text{IL} = \text{SPL} - 0.0168$$

The characteristics of the human ear are such that in practice it is never worthwhile to work to an accuracy of more than one dB, and all practical results should always be rounded off to the nearest decibel. So we see that for all practical measurements on the decibel scale, the intensity level is numerically equal to the sound pressure level.

Let us now consider the addition of two sounds of equal sound pressure (SPL). Then the SPL of the total can be written

$$\text{SPL} = 10 \log\left(\frac{p_t^2}{p_0^2}\right) = 10 \log\left(\frac{2p_1^2}{p_0^2}\right) \text{dB}$$

$$= 10 \log\left(\frac{p_1^2}{p_0^2}\right) + 10 \log 2$$

$$\text{SPL} = (\text{SPL})_1 + 3 \text{ dB}$$

This shows that we cannot simply add decibels arithmetically. In general, if two quantities x and y are measured on the decibel scale with respect to a reference value a, we have by definition

$$B_x = 10 \log (x/a)$$

$$B_y = 10 \log (y/a)$$

$$B = 10 \log (x/a + y/a)$$

Hence

$$B = 10 \log (y/a)(1 + x/y)$$

$$= 10 \log (y/a) + 10 \log (1 + x/y)$$

$$= B_y + 10 \log (1 + x/y) \tag{2.7}$$

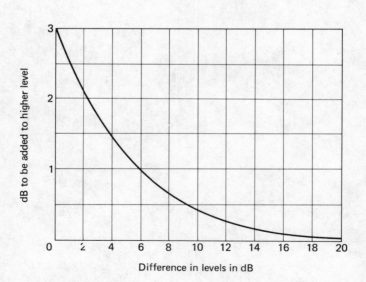

2.4 The addition of decibels

But

$$B_x - B_y = 10 \log (x/a) - 10 \log (y/a)$$
$$= 10 \log (x/y)$$
$$\therefore \; x/y = \text{antilog} \; [(B_x - B_y)/10] \qquad (2.8)$$

Substituting eqn 2.8 in eqn 2.7, we have the rule for adding decibel levels,

$$B = B_y + 10 \log \{1 + \text{antilog} \; [(B_x - B_y)/10] \} \qquad (2.9)$$

This is a cumbersome expression to use, and charts have been prepared which enable the addition to be done without having to look up logs and antilogs. Fig. 2.4 is a decibel addition curve.

An incidental advantage of the decibel scale is that it does correspond roughly to our subjective response to physical stimuli. This is expressed in the Weber—Fechner Law which states that the response varies as the logarithm of the stimulus.

3   Spherical waves

3.1   Plane waves and spherical waves

The method which we have used for arriving at some of the equations derived so far applies strictly speaking only to **plane waves**.

These are waves in which the wave front — an imaginary surface joining points of constant particle displacement, s — forms a plane of constant area perpendicular to the direction of travel of the wave.

We made an implicit assumption that this was the kind of wave we were dealing with by making s a function of the direction x but not of the directions y and z. In this case the particle velocity u and the acoustic pressure p also appear as functions of x only.

Sometimes waves originate at a point source and spread out in all directions from the source. In this case the displacement, particle velocity and pressure are no longer constant over any plane. Instead they are constant over the surface of a spherical shell surrounding the source. This type of wave is known as a **spherical wave.**

Many of the results for plane waves also apply to such spherical waves, but some need to be modified to take account of the spreading out of the wave. For the plane wave, in the absence of any material to absorb the sound energy, the pressure and intensity are independent of the distance travelled by the wave. For a spherical wave the pressure varies inversely with distance, and intensity, being proportional to pressure squared, varies inversely with the square of the distance.

3.2   Analysis of spherical waves

The method of dealing with a plane wave can be extended from one dimension to three to cover the case of spherical waves. It is convenient to use spherical co-ordinates, so that we will be able to make full use of the spherical symmetry. In spherical co-ordinates, the position of a point in space P is given by the distance r from the origin O, the angle θ between the line OP and the z-axis, and the angle ψ between the x-axis and the plane POz. The relationship between spherical and rectangular co-ordinates is illustrated in fig. 3.1 which shows that

$$x = r \sin \theta \cos \psi \qquad\qquad r = (x^2 + y^2 + z^2)^{1/2}$$

$$y = r \sin \theta \sin \psi \qquad\qquad \psi = \arctan (y/x)$$

$$z = r \cos \psi \qquad\qquad \theta = \arctan (x^2 + y^2)^{1/2}/2$$

The spherical wave equation is given in eqn 3.8. Its full derivation, given here, although not difficult, is a little tedious and may be omitted by those who are prepared to take the result on trust.

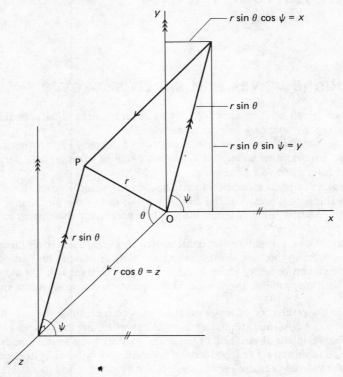

3.1 Spherical polar co-ordinates

Fig. 3.2 shows an infinitesimal element with sides of lengths δr, $r\,\delta\theta$ and $r\sin\theta\;d\psi$. It is bounded by three pairs of opposite faces with areas $r\,\delta r\,\delta\theta$, $r\sin\theta\,\delta\psi\,\delta r$ and $r^2\sin\theta\,\delta\theta\,\delta\psi$, and contains a volume $r^2\sin\theta\,\delta\theta\,\delta\psi\,\delta r$. In deriving the wave equation we first found an expression for the change in volume of such an element due to displacement s. When there is symmetry about the origin, s is in the same direction as r, and the decrease in volume is

$$\left(s-\frac{\partial s}{\partial r}\frac{\delta r}{2}\right)\left(r-\frac{\delta r}{2}\right)^2\sin\theta\,\delta\theta\,\delta\psi-\left(s+\frac{\partial s}{\partial r}\frac{\delta r}{2}\right)\left(r+\frac{\delta r}{2}\right)^2\sin\theta\,\delta\theta\,\delta\psi$$

$$=\left[r^2-r\,\delta r+\left(\frac{\delta r}{2}\right)^2\right]\left(s-\frac{\partial s}{\partial r}\frac{\delta r}{2}\right)\sin\theta\,\delta\theta\,\delta\psi-\left[r^2+r\,\delta r+\left(\frac{\delta r}{2}\right)^2\right]$$

$$\times\left(s+\frac{\partial s}{\partial r}\frac{\delta r}{2}\right)\sin\theta\,\delta\theta\,\delta\psi$$

$$=\left[-r^2\frac{\partial s}{\partial r}\,\delta r-2rs\,\delta r-\frac{\partial s}{\partial r}\frac{(\delta r)^3}{4}\right]\sin\theta\,\delta\theta\,\delta\psi$$

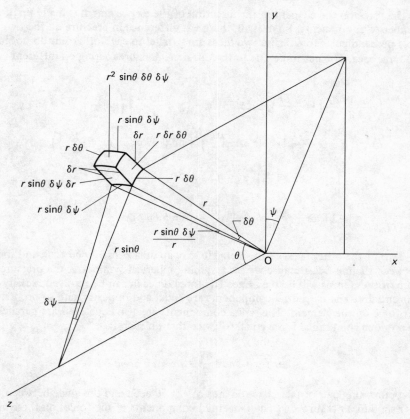

3.2 An infinitesimal element in spherical co-ordinates

Dividing by the original volume $r^2 \sin\theta \,\delta\theta \,\delta\psi \,\delta r$ and taking the limit as δr tends to zero, we have

$$\text{volumetric strain} = -\frac{\partial s}{\partial r} - \frac{2s}{r}$$

As before

$$E = \frac{\text{stress}}{\text{strain}}$$

and so

$$p = -E\frac{\partial s}{\partial r} - 2\frac{Es}{r}$$

This can be written more concisely as

$$p = -E\frac{1}{r}\frac{\partial}{\partial r^2}(r^2 s) \tag{3.1}$$

The force on the element in the direction of the displacement is made up of a number of components. First of all, there is a difference in pressure on the two faces perpendicular to r. These two faces are parallel to each other but do not have the same area. The net force due to the different pressures acting on different areas is

$$\left(p - \frac{\partial p}{\partial r}\frac{\delta r}{2}\right)\left(r - \frac{\delta r}{2}\right)^2 \sin\theta\,\delta\theta\,\delta\psi - \left(p + \frac{\partial p}{\partial r}\frac{\delta r}{2}\right)\left(r + \frac{\delta r}{2}\right)^2 \sin\theta\,\delta\theta\,\delta\psi$$

$$= \left(p - \frac{\partial p}{\partial r}\frac{\delta r}{2}\right)\left[r^2 - r\,\delta r + \left(\frac{\delta r}{2}\right)^2\right]\sin\theta\,\delta\theta\,\delta\psi - \left(p + \frac{\partial p}{\partial r}\frac{\delta r}{2}\right)$$

$$\times \left[r^2 + r\,\delta r + \left(\frac{\delta r}{2}\right)^2\right]\sin\theta\,\delta\theta\,\delta\psi$$

$$= \left[-2pr\,\delta r - r^2\frac{\partial p}{\partial r}\delta r - \frac{\partial p}{\partial r}\frac{(\delta r)^3}{4}\right]\sin\theta\,\delta\theta\,\delta\psi \tag{3.2}$$

Two of the other faces of the element have an area $r\sin\theta\,\delta r\,\delta\theta$ each and the angle between them is $\delta\psi$. Because we are assuming spherical symmetry, the pressure on each of these faces will be the same. The force on each can be resolved radially and tangentially. The tangential components are equal and opposite, and so produce no net force on the element. The radial components are also equal, but are parallel and so produce a net radial force equal to twice that on each face,

$$2pr\sin\theta\,\delta r\,\delta\theta\,\frac{\delta\psi}{2} = pr\sin\theta\,\delta r\,\delta\theta\,\delta\psi \tag{3.3}$$

The remaining pair of faces have an area of $r\,\delta r\,\delta\theta$ each and the angle between them is $(r\sin\theta\,\delta\psi)/r = \sin\theta\,\delta\psi$. The tangential components of the forces on these areas again cancel, and the radial components give a net force

$$2pr\,\delta r\,\delta\theta\,(\sin\theta\,\delta\psi/2) = pr\sin\theta\,\delta r\,\delta\theta\,\delta\psi \tag{3.4}$$

The total radial force is the sum of eqns 3.2, 3.3 and 3.4 giving

$$\text{radial force} = \left[-2pr\,\delta r - r^2\frac{\partial p}{\partial r}\delta r - \frac{\partial p}{\partial r}\frac{(\delta r)^3}{4} + pr\,\delta r + pr\,\delta r\right]\sin\theta\,\delta\theta\,\delta\psi$$

$$= -\left[r^2\frac{\partial p}{\partial r}\delta r + \frac{\partial p}{\partial r}\frac{(\delta r)^3}{4}\right]\sin\theta\,\delta\theta\,\delta\psi$$

The mass of the element is $pr^2\sin\theta\,\delta\theta\,\delta\psi\,\delta r$, and the radial acceleration is $\partial^2 s/\partial t^2$. Therefore

$$-\left[r^2\frac{\partial p}{\partial r} + \frac{\partial p}{\partial r}\frac{(\delta r)^2}{4}\right]\sin\theta\,\delta\theta\,\delta\psi\,\delta r = \rho r^2\sin\theta\,\delta\theta\,\delta\psi\,\delta r\,\frac{\partial^2 s}{\partial t^2}$$

Again taking the limit δr tends to zero, we are left with

$$\frac{-\partial p}{\partial r} = \rho\frac{\partial^2 s}{\partial t^2} \tag{3.5}$$

We now have two differential equations, eqns 3.1 and 3.5, containing p and s. To find an equation for p alone we must eliminate s between the two equations. To do so we differentiate eqn 3.1 twice with respect to t, remembering that r is independent of t. This gives

$$\frac{\partial^2 p}{\partial t^2} = -\frac{E}{r^2} \frac{\partial}{\partial r} \frac{\partial^2}{\partial t^2} (r^2 s) \tag{3.6}$$

Again, since r is independent of t, eqn 3.5 can be multiplied by r^2 and re-written

$$-r^2 \frac{\partial p}{\partial r} = \rho \frac{\partial^2}{\partial t^2} (r^2 s)$$

Differentiating with respect to r gives

$$\frac{\partial}{\partial r} \left(r^2 \frac{\partial p}{\partial r} \right) = -\rho \frac{\partial}{\partial r} \frac{\partial^2}{\partial t^2} (r^2 s) \tag{3.7}$$

Comparing eqns 3.6 and 3.7 gives

$$\frac{\partial^2 p}{\partial t^2} = \frac{E}{\rho} \frac{1}{r^2} \frac{\partial}{\partial r} \left(r^2 \frac{\partial p}{\partial r} \right)$$

By means of a little re-arranging we can make this look more like the wave equation, eqn 1.1:

$$r \frac{\partial^2 p}{\partial t^2} = \frac{E}{\rho} \frac{1}{r} \left[2r \frac{\partial p}{\partial r} + r^2 \frac{\partial^2 p}{\partial r^2} \right]$$

$$\therefore \frac{\partial^2}{\partial t^2} (rp) = \frac{E}{\rho} r \left[\frac{\partial^2 p}{\partial r^2} + \frac{2}{r} \frac{\partial p}{\partial r} \right]$$

$$\therefore \frac{\partial^2}{\partial t^2} (rp) = \frac{E}{\rho} \frac{\partial^2}{\partial r^2} (rp) \tag{3.8}$$

The last step follows from

$$\frac{\partial}{\partial r} (rp) = p + r \frac{\partial p}{\partial r}$$

$$\frac{\partial^2}{\partial r^2} (rp) = \frac{\partial p}{\partial r} + r \frac{\partial^2 p}{\partial r^2} + \frac{\partial p}{\partial r} = r \left(\frac{\partial^2 p}{\partial r^2} + \frac{2}{r} \frac{\partial p}{\partial r} \right)$$

Eqn 3.8 is the spherical wave equivalent of eqn 1.1. It is a wave equation for the quantity (rp) and we may write the solution as

$$rp = A \cos 2\pi (ft - r/\lambda)$$

or
$$p = (A/r) \cos 2\pi (ft - r/\lambda) \tag{3.9}$$

We see straight away that a spherical wave behaves in general just like a plane wave, except that p varies inversely with r. Consideration of speed of propagation, frequency and wavelength are exactly as discussed for plane waves. The calculation of r.m.s. pressures and the addition of pressures are exactly the same.

3.3 Pressure and particle velocity

The relationship between acoustic pressure p and particle velocity u is no longer quite the same. The displacement s takes place in the radial direction, so the force equation becomes

$$\frac{\partial p}{\partial r} = -\rho \frac{\partial^2 s}{\partial t^2}$$

From eqn 3.9 this may be written

$$\frac{\partial p}{\partial r} = -\frac{A}{r^2} \cos 2\pi (ft - r/\lambda) - \frac{2\pi A}{r\lambda} \sin 2\pi (ft - r/\lambda)$$

Now

$$u = \frac{\partial s}{\partial t} = -\frac{1}{\rho} \int \frac{\partial p}{\partial r}\, dt$$

$$= \int \frac{A}{r^2 \rho} \cos 2\pi (ft - r/\lambda)\, dt + \int \frac{2\pi A}{r\lambda\rho} \sin 2\pi (ft - r/\lambda)\, dt$$

$$= \frac{A}{2\pi fr^2 \rho} \sin 2\pi (ft - r/\lambda) - \frac{2\pi A}{2\pi fr\lambda\rho} \cos 2\pi (ft - r/\lambda)$$

$$u = \frac{A}{2\pi fr^2 \rho} \left\{ \sin 2\pi (ft - r/\lambda) - 2\pi \frac{r}{\lambda} \cos 2\pi (ft - r/\lambda) \right\}$$

We see that p and u are no longer in phase. The phase angle between them is $\theta = \arctan 1/(2\pi r/\lambda)$ and

$$u = \frac{A}{2\pi fr^2 \rho} [1 + (2\pi r/\lambda)^2]^{\frac{1}{2}} \left\{ \frac{1}{[1 + (2\pi r/\lambda)^2]^{\frac{1}{2}}} \sin 2\pi (ft - r/\lambda) \right.$$

$$\left. - \frac{(2\pi r/\lambda)}{[1 + (2\pi r/\lambda)^2]^{\frac{1}{2}}} \cos 2\pi (ft - r/\lambda) \right\}$$

$$= \frac{A [1 + (2\pi r/\lambda)^2]^{\frac{1}{2}}}{2\pi fr^2 \rho} \left\{ \sin \theta \sin 2\pi (ft - r/\lambda) - \cos \theta \cos 2\pi (ft - r/\lambda) \right\}$$

$$= \frac{A}{r} \frac{[1 + (2\pi r/\lambda)^2]^{\frac{1}{2}}}{\rho c (2\pi r/\lambda)} \cos 2\pi (ft - r/\lambda - \theta/2\pi) \qquad (3.10)$$

(An alternative derivation of this equation is given in Appendix 1).

Apart from the phase angle, the numerical value of the ratio between p and u contains a factor $(1 + (2\pi r/\lambda)^2)^{\frac{1}{2}}/(2\pi r/\lambda)$ for spherical waves which does not appear in the corresponding expression for plane waves. Let us see how important this is in practice by calculating what value of r/λ is needed to make this factor differ from unity by more than 10%.

We require

$$\frac{(1 + (2\pi r/\lambda)^2)^{\frac{1}{2}}}{2\pi r/\lambda} > 1.1$$

$$\therefore \quad 1 + (2\pi r/\lambda)^2 > 1.21 \, (2\pi r/\lambda)^2$$

$$\therefore \quad 0.21 \, (2\pi r/\lambda)^2 < 1$$

$$\therefore \quad 0.46 \, (2\pi r/\lambda) < 1$$

$$\therefore \quad 2\pi r/\lambda < 2.18$$

$$\therefore \quad r/\lambda < 0.35$$

If r is less than about one third of a wavelength, the difference between the value for spherical and plane waves is more than 10%. For values of r greater than this the plane wave equation can be used without great error. The greatest value of λ corresponds to the lowest value of f and taking 50 Hz as the lowest frequency of practical importance, gives $\lambda = 6.73$ metres. So we can say that at all distances more than about 2.5 metres from the source, the plane wave equation can be used as a good approximation even if we know the wave is in fact spherical.

3.4   Energy density

We can use eqns 3.9 and 3.10 to find the energy density per unit volume, in a similar way to that used for plane waves. As before

$$\mathcal{E} = E/V_0 = p^2/2\rho c^2 + \rho u^2/2$$

$$= \frac{1}{2}\left\{ \frac{A^2}{c^2\rho r^2} \cos^2 2\pi \, (ft - r/\lambda) + \frac{A^2}{c^2\rho r^2} \frac{1 + (2\pi r/\lambda)^2}{(2\pi r/\lambda)^2} \cos^2 2\pi \, (ft - r/\lambda - \theta/2\pi) \right\}$$

$$= \frac{A^2}{4c^2\rho r^2} \left\{ 1 + \cos 4\pi \, (ft - r/\lambda) + \frac{1 + (2\pi r/\lambda)^2}{(2\pi r/\lambda)^2} \right.$$

$$\left. + \frac{1 + (2\pi r/\lambda)^2}{(2\pi r/\lambda)^2} \cos 4\pi \times (ft - r/\lambda - \theta/2\pi) \right\} \tag{3.11}$$

Taking the time average,

$$\mathcal{E}_{\text{mean}} = \frac{1}{4c^2\rho} \left(\frac{A}{r}\right)^2 \left[1 + \frac{1 + (2\pi r/\lambda)^2}{(2\pi r/\lambda)^2} \right]$$

But comparing eqns 3.9, 2.1 and 2.2 we see that for the spherical wave we are considering

$$A^2/2r^2 = p_{\text{r.m.s.}}^2$$

So

$$\mathcal{E}_{\text{mean}} = \frac{p_{\text{r.m.s.}}^2}{2c^2\rho} \left[\frac{1 + 2(2\pi r/\lambda)^2}{(2\pi r/\lambda)^2} \right]$$

$$= \frac{p_{\text{r.m.s.}}^2}{c^2\rho} \left[1 + \frac{1}{2(2\pi r/\lambda)^2} \right] \tag{3.12}$$

Let us again consider the error introduced by using eqn 2.6 as an approximation to eqn 3.12. If the last term in brackets is to be less than 0.1, then $2(2\pi r/\lambda)^2$ must be more than 10, which makes $(2\pi r/\lambda)$ more than 2.24 and r/λ must be greater than 0.36. So here again the difference between plane and spherical waves is in practice negligible except within 2 or 3 metres of the source.

3.5 ⎸ Intensity

To find the intensity of plane waves we considered the energy in a volume $\delta x\,\delta y\,\delta z$ and the rate at which this energy flowed out of the $\delta y\,\delta z$ face in the δx direction. We found the energy flowing out of the volume in time δt by making $\delta x = c\,\delta t$. A different approach is needed for spherical waves.

Consider the rate at which a small element of the fluid does work on the neighbouring fluid. This rate is the power, and is equal to the force exerted times the particle velocity. The force is simply the pressure times the elemental area, so the intensity, or power per unit area, is the product of pressure and particle velocity. The mean intensity is obtained by taking the time average of this product. For a spherical wave

$$I_{\text{mean}} = \int_0^T pu \, \mathrm{d}t/T$$

$$= \frac{1}{T} \int_0^T \frac{A}{r} \cos 2\pi \left(ft - \frac{r}{\lambda} \right) \left\{ \frac{A}{2\pi fr^2 \rho} \sin 2\pi \left(ft - \frac{r}{\lambda} \right) \right.$$

$$\left. - \frac{A}{fr\lambda\rho} \cos 2\pi \times \left(ft - \frac{r}{\lambda} \right) \right\} \, \mathrm{d}t \tag{3.13}$$

Now

$$\int \cos \theta \, \sin \theta \, \mathrm{d}\theta = \int \sin \theta \, \mathrm{d}(\sin \theta) = |\tfrac{1}{2} \sin^2\theta| = 0 \quad \text{over a whole cycle}$$

and

$$\int \cos^2 \theta \, \mathrm{d}\theta = \tfrac{1}{2} \int \mathrm{d}\theta + \tfrac{1}{2} \int \cos 2\theta \, \mathrm{d}\theta = \tfrac{1}{2} \int \mathrm{d}\theta \quad \text{over a whole cycle.}$$

The only part of the product pu which contributes to the steady flow of energy is due to the component of u which is in phase with p. The component of u which is out of phase with p gives a flow of energy in one direction during one part of the cycle, and an exactly equal flow of energy in the opposite direction during the next part of the cycle.

In the long run the net flow of energy due to this component is nil. It does however contribute to the energy density. If we calculate the intensity from the energy density, as we did for a plane wave, we must use only the first term in the brackets of eqn 3.12, because the second term is due to the out of phase components of p and u.

The expression for the intensity of spherical waves is then the same as for plane waves, as may also be seen by completing the integration of eqn 3.13.

$$I_{\text{mean}} = \frac{1}{T} \left\{ \frac{A^2}{r^2 f \lambda \rho} \right\} \int_0^T \tfrac{1}{2} \, dt = \tfrac{1}{2} \left(\frac{A}{r} \right)^2 \frac{1}{\rho c}$$

$$= \frac{p_{\text{r.m.s.}}^2}{\rho c}$$

The relationship between intensity and pressure is one of the most useful encountered in acoustics and it is important to remember that it is the same for plane and spherical waves.

3.6 Cylindrical waves

One can also consider a wave which diverges in two dimensions. Such a wave could, for example, originate at the axis of a cylinder and the wave fronts would be a series of concentric cylinders. For this problem it is convenient to use polar co-ordinates, which are illustrated in fig. 3.5, and for which

$$x = r \cos \theta \qquad r = (x^2 + y^2)^{\frac{1}{2}}$$

$$y = r \sin \theta \qquad \theta = \arctan (y/x)$$

$$z = z$$

Fig. 3.4 shows an infinitesimal element with sides of length δr, $r \, \delta \theta$ and δz. Its faces have areas of $r \, \delta \theta \, \delta r$, $\delta r \, \delta z$ and $r \, \delta z \, \delta \theta$; it has a volume of $r \, \delta \theta \, \delta z \, \delta r$. The decrease in volume due to a displacement s in the r direction is

$$\left(s - \frac{\partial s}{\partial r} \frac{\delta r}{2} \right) \left(r - \frac{\delta r}{2} \right) \delta \theta \, \delta z - \left(s + \frac{\partial s}{\partial r} \frac{\delta r}{2} \right) \left(r + \frac{\delta r}{2} \right) \delta \theta \, \delta z = \left[-s \, \delta r - r \frac{\partial s}{\partial r} \delta r \right] \delta \theta \, \delta z$$

Dividing by the original volume gives

$$\text{volumetric strain} = -\frac{\partial s}{\partial r} - \frac{s}{r}$$

$$\therefore p = -E \frac{\partial s}{\partial r} - E \frac{s}{r}$$

which can be written more conveniently as

$$p = -E \frac{1}{r} \frac{\partial}{\partial r} (rs) \tag{3.13}$$

The force on the element in the direction of the displacement has a component due to different pressures acting on different areas perpendicular to r. This component is

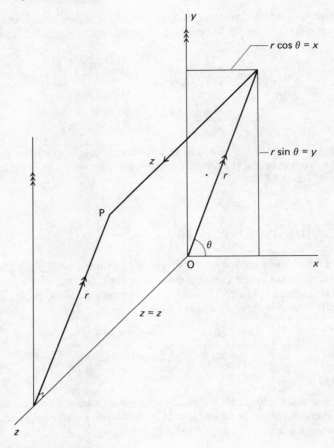

3.3 Cylindrical polar co-ordinates

$$\left(p - \frac{\partial p}{\partial r}\frac{\delta r}{2}\right)\left(r - \frac{\delta r}{2}\right)\delta z\,\delta\theta - \left(p + \frac{\partial p}{\partial r}\frac{\delta r}{2}\right)\left(r + \frac{\delta r}{2}\right)\delta z\,\delta\theta = \left(-p\,\delta r - r\frac{\partial p}{\partial r}\delta r\right)\delta\theta\,\delta z$$

$$(3.14)$$

There is also a component in the radial direction due to the pressure on the faces $\delta r\,\delta z$. This component is

$$2p\,\delta r\,\delta z\,\delta\theta/2 = p\,\delta r\,\delta z\,\delta\theta \tag{3.15}$$

The faces $r\,\delta\theta\,\delta r$ are parallel to each other and at right angles to the z-axis. Since we are assuming that the pressure is constant in the z direction, the pressure on these two faces cancel out. The total radial force is then the sum of eqns 3.14 and 3.15,

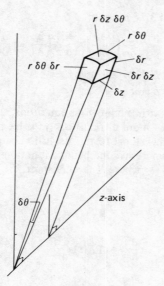

3.4 An infinitesimal element in cylindrical co-ordinates

$$\text{radial force} = -p\,\delta r\,\delta z\,\delta\theta - r\frac{\partial p}{\partial r}\delta r\,\delta z\,\delta\theta + p\,\delta r\,\delta z\,\delta\theta$$

$$= -\frac{\partial p}{\partial r}r\,\delta r\,\delta z\,\delta\theta$$

The mass of the element is $\rho r\,\delta\theta\,\delta r\,\delta z$ and the radial acceleration is $\partial^2 s/\partial t^2$. Therefore

$$-\frac{\partial p}{\partial r} = \rho\frac{\partial^2 s}{\partial t^2} \tag{3.16}$$

We must now eliminate s between eqns 3.13 and 3.16. We use the same technique as before, and have from eqn 3.13

$$\frac{\partial^2 p}{\partial t^2} = -\frac{E}{r}\frac{\partial}{\partial r}\frac{\partial^2}{\partial t^2}(rs) \tag{3.17}$$

and from eqn 3.16

$$\frac{r}{\partial r}\frac{\partial p}{\partial r} = -\rho\frac{\partial^2}{\partial t^2}(rs)$$

$$\therefore \frac{\partial}{\partial r}\left(r\frac{\partial p}{\partial r}\right) = -\rho\frac{\partial}{\partial r}\frac{\partial^2}{\partial t^2}(rs) \tag{3.18}$$

Combining eqns 3.17 and 3.18 leads to

$$\frac{\partial^2 p}{\partial t^2} = \frac{E}{\rho} \frac{1}{r} \frac{\partial}{\partial r} \left(r \frac{\partial p}{\partial r} \right) = \frac{E}{\rho} \frac{1}{r} \left(\frac{\partial p}{\partial r} + r \frac{\partial^2 p}{\partial r^2} \right)$$

$$= \frac{E}{\rho} \left(\frac{\partial^2 p}{\partial r^2} + \frac{1}{r} \frac{\partial p}{\partial r} \right) \tag{3.19}$$

In spite of the similarity with the previous equations, this two-dimensional version of the wave equation is a more difficult one to solve than either the one-dimensional or the three-dimensional form. A solution can be found by assuming that p is a product of two functions, one being a function only of r and the other only of t. Thus $p = AR(r)T(t)$ where A is a constant.

Differentiating gives

$$\frac{\partial p}{\partial r} = AT(t) \frac{dR}{dr}$$

$$\frac{\partial^2 p}{\partial r^2} = AT(t) \frac{d^2 R}{dr^2}$$

and $$\frac{\partial^2 p}{\partial t^2} = AR(r) \frac{d^2 T}{dt^2}$$

Substituting in eqn 3.19 gives

$$AR(r) \frac{d^2 T}{dt^2} = \frac{E}{\rho} AT(t) \left\{ \frac{d^2 R}{dr^2} + \frac{1}{r} \frac{dR}{dr} \right\}$$

$$\therefore \frac{\rho}{E} \frac{1}{T(t)} \frac{d^2 T}{dt^2} = \frac{1}{R(r)} \left\{ \frac{d^2 R}{dr^2} + \frac{1}{r} \frac{dR}{dr} \right\}$$

The left hand side is a function only of t while the right hand side is a function only of r; they can be equal to each other for all values of r and t only if they are both equal to some constant. Let us call this constant $-a^2$.

Then

$$\frac{d^2 T}{dt^2} = -a^2 \frac{E}{\rho} T(t)$$

which gives

$$T = \cos \left[a\sqrt{(E/\rho)} t \right]$$

We have seen that $\sqrt{(E/\rho)}$ is the velocity of sound c. The constant a must have the dimensions of 1/length to make the cosine term dimensionless, and we can guess that it will turn out to be the reciprocal of the wavelength, so that we may write

$$T = \cos (c/\lambda) t$$

We also have

$$\frac{d^2 R}{dr^2} + \frac{1}{r} \frac{dR}{dr} = -a^2 R$$

or

$$\frac{d^2R}{dr^2} + \frac{1}{r}\frac{dR}{dr} + \frac{R}{\lambda^2} = 0$$

This is Bessel's equation of zero order, and the solution is written as

$$R = J_0 \, (r/\lambda)$$

where J_0 is a Bessel function of zero order.
The complete solution of eqn 3.19 is then

$$p = AJ_0 \, (r/\lambda) \cos \, (ct/\lambda)$$

Values of J_0 have been tabulated in the same way as values of cos and sin, but the tables are not so readily available. Moreover, practical problems are more likely to involve plane or spherical waves than cylindrical waves. We shall therefore not take the theory any further in terms of Bessel functions, but shall instead see whether the plane wave equations can be used as an approximation, as they could for spherical waves.

As r/λ increases, an approximate equation can be written for $J_0 \, (r/\lambda)$.

$$J_0 \, (r/\lambda) \approx \left(\frac{2}{\pi r/\lambda}\right)^{\frac{1}{2}} \cos \, [(r/\lambda) - (\pi/4)]$$

Substituting numerical values from the published tables of J_0 and cos shows that if r/λ is greater than 1, the error introduced by this approximation is not more than about 5 per cent. This is quite good enough for all acoustic calculations, and allows us to express the acoustic pressure as

$$p = A \, (2\lambda/\pi)^{\frac{1}{2}} (1/\sqrt{r}) \cos \, [(r/\lambda) - (\pi/4)] \, \cos \, (ct/\lambda)$$

$$= A \, (\lambda/2\pi)^{\frac{1}{2}} (1/\sqrt{r}) \cos \, [(r/\lambda) - ct/\lambda) - (\pi/4)] \, + A \, (\lambda/2\pi)^{\frac{1}{2}} (1\sqrt{r})$$

$$\times \cos \, [(r/\lambda) + (ct/\lambda) - (\pi/4)]$$

The first term will be recognized as a wave travelling in the positive r direction with velocity c, wavelength λ, and an amplitude varying inversely with $\sqrt{r}$. The second term is an identical wave travelling in the opposite direction. The phase angle $\pi/4$ can be eliminated by choosing a different point in time from which to measure t. The full solution of the plane wave equation also gives two waves travelling in opposite directions. We considered only one of the two travelling waves, because the one travelling in the opposite direction does not add anything to the physical significance of the solution.

3.7 General approximation

We see that a cylindrical wave can, to an acceptable approximation, be treated as a plane wave whose pressure amplitude diminishes as $1/\sqrt{r}$. The energy and intensity, being proportional to p^2, will then diminish as $1/r$. This is exactly what we would expect from physical considerations. The wave front of a cylindrical wave is the surface of a cylinder; the area of the wave front increases as the radius of the cylinder is increased. If a constant amount of energy is being propagated, the intensity must fall as the radius increases so that the product of intensity and area remains constant. Thus the intensity will be inversely proportional to the area and so to the radius.

In the same way, a spherical wave front is the surface of a sphere, where area is proportional to the square of the radius. The intensity is then inversely proportional to the square of the radius.

The way in which the intensity varies with distance can always be found by remembering that intensity times area, taken over the whole of the wave front, must remain constant as the wave front progresses. This is inherent in the concept of intensity, and remains the basic physical idea no matter how elaborate the mathematics becomes.

In other respects all waves approximate to plane waves; the larger the distance from the origin, the better is the approximation. Again, this agrees with our physical ideas, for at large distances from the centre of curvature, the curvature is small and may be ignored. In acoustic calculations the approximation is almost always valid. Nevertheless some pitfalls can be avoided by remembering that this is an approximation.

The most important result which remains valid for all waves is that the intensity is proportional to the square of the pressure.

4 Loudness

4.1 Subjective measurements

Most people are used to the idea that although quantities such as temperature, humidity, and air velocity can be measured with a considerable degree of accuracy, it is far from simple to find any very precise correlation between such physical measurements and any one individual's personal feeling of comfort. It is everyone's experience that no two people agree on what is pleasant and what is disagreeable, so that even if any acceptable correlation were found for one person, it would be of little practical use. Human beings react to conditions around them in a highly subjective manner and the subjective effect of environmental conditions has not yet been reduced to a simple numerical scale.

The same problems arise in interpreting sound measurements and in particular in deciding how the amplitudes and frequencies we have been discussing so far are related to the feeling of loudness. In spite of the difficulty of all subjective measurements, it is necessary to have some means of estimating how people are likely to react to a sound of known physical qualities. The whole art of applied acoustics would be pointless without some such criteria.

The basis of all sound criteria is a statistical investigation into the reactions of a large number of people. Suppose one person sits in a room in which a loudspeaker is generating a known acoustic pressure. If the pressure is steadily reduced, there will come a point when the person being tested will say 'I can no longer hear anything'. The lower limit of his hearing has been reached. Another person will have a slightly different limit. The limit for any one person may depend on his mood at the time he is being tested, and will certainly depend to some extent on his general health, so it may well vary from day to day. If a large number of people are tested in such a way, a range of values will be obtained for the lowest acoustic pressure which they can hear. The average of this range can be taken as the practical limit of audibility.

By means of such investigations, a considerable amount of information has been obtained about the way the human ear responds to acoustic vibrations. To carry out these tests does not require any knowledge about the structure of the ear, and the results can be described, understood, and above all used in the practical application of acoustics, without having to discuss the detailed anatomy and physiology of the ear. What is important to realize is that there is a statistical variation in the response of people having normal hearing, and that the natural variation between different people has been smoothed out in the published data.

The three main subjective qualities of a sound are its **loudness**, **pitch**, and **timbre**; of these by far the most important in acoustic engineering is loudness. The loudness corresponds largely to the intensity and sound pressure level, but it is also a function of frequency. If on a sheet of graph paper we plot sound pressure level along the y-axis, and frequency along the x-axis, then any point on the graph represents a pure sound of one frequency and one intensity. Suppose a sound of some other frequency and intensity appears equally loud to a large number of people, or to the statistically average listener, we now have two points which can be described as being equally loud. Let us draw a line joining all points which appear to the average listener to be equally loud as the first sound. This line is called an **equal loudness contour**.

The process can be repeated for a sound of some other loudness to give another equal loudness contour. We can go on to produce any number of such curves, and so build up a chart of equal loudness contours, such as is reproduced in fig. 4.1.

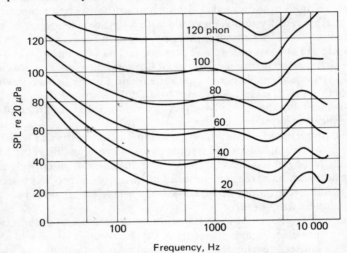

4.1 Equal loudness contours (After Robinson and Dadson, *British Journal of Applied Physics*, 7, 1956)

The apparent precision of these curves should not make us forget that they are based on a large number of individual judgements.

The next step is to give each loudness contour a label by attaching a numerical value to it. The value which has been chosen is the SPL at the point where the curve crosses or touches the 1000 Hz ordinate. This number is called the **loudness level in phons** of that particular equal loudness contour. All points lying on the same contour have the same loudness expressed in phons. The loudness level in phons of any sound can be defined as the SPL of a pure tone of 1000 Hz which sounds equally loud to the average listener. This definition makes the phon scale logarithmic in exactly the same way as the decibel scale of SPL.

It has been found that the ear responds to acoustic pressure ranging from about 10^{-4} to about 10^3 microbars and to frequencies from about 20 Hz to about 15 000 Hz. The very large range of pressures in which we are therefore interested is another very good reason for using the logarithmic decibel scale. The limit below which we cannot hear is called the **threshold of audibility**, and the other extreme, the limit above which sound becomes painful, is called the **threshold of feeling**. Both these thresholds vary with the frequency of the sound. They also vary with the listener's age, particularly at frequencies above 2000 Hz. This can be an important point in providing adequate silencing in practice, for there is no point in spending money and effort on more silencing than is needed, and how

much is needed may, in certain circumstances, depend on the age of the persons
for whose benefit the silencing is being provided.

The threshold of audibility and the thresholds of feeling and of pain can be
plotted on an SPL–frequency graph as in fig. 4.2. The limits form a closed curve,
and all points within the enclosed area represent sounds which can be heard by
the human ear. All points outside the closed curve represent inaudible vibrations.

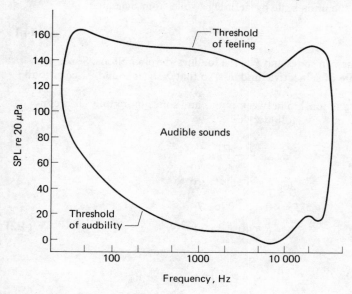

4.2 The limits of audibility

Although the ear responds to frequencies from 20 to 15 000 Hz, it is not
normally necessary to deal with quite such a wide frequency range. Ordinary
speech covers a range from a little below 200 to something above 10 000 Hz,
but only the range from 1000 to 2500 Hz is necessary for the speech to be
understood. Telephone systems usually transmit a frequency range from about
500 to 3000 Hz. The lower frequencies in this range are needed only to make the
speech sound fairly natural. Without them it could be understood, but the speaker's
voice would be difficult to recognize. The reproduction of music requires a wider
frequency band. The piano keyboard, for example, extends from 27.5 to 3250 Hz.
It is rarely necessary, in acoustic calculations, to consider frequencies above
4000 Hz, and usually it is enough to use 62.5 Hz for the lowest band mid-frequency
and 2000 Hz for the highest band mid-frequency.

4.2   Numerical values

Although we have attached a numerical value to each curve and called it loudness
level in phons, we have still no means of expressing how much louder one sound

is than another. This can be done by once again testing a large number of people. The result of such experiments has shown that the average person will say that the loudness of a sound has doubled when the level has increased by approximately 10 phons.

It would in some ways be convenient if we had a scale in which the numerical value doubled for a doubling of the subjective feeling of loudness. Such a scale can be derived from the phon scale by defining a **sone** from the equation

$$S = 2^{(P-40)/10} \qquad (4.1)$$

where S is the loudness in sones and P is the loudness level in phons. Sones are intended as a measure of subjective loudness, so that 2 sones sound twice as loud as 1 sone.

To illustrate the relationship between phons and sones more fully, it is convenient to re-write eqn 4.1 as follows:

$$\log S = \frac{P-40}{10} \log 2$$

$$\therefore \quad \log S = 0.3\,(P-40)$$

$$\therefore P - 40 = 33.3 \log S$$

$$\therefore \qquad P = 40 + 33.3 \log S \qquad (4.2)$$

If we now write

$$P_2 = P_1 + 10$$

then

$$P_2 = 40 + 33.3 \log S_2$$

$$= 50 + 33.3 \log S_1$$

$$\therefore \ 33.3 \log S_2 - 33.3 \log S_1 = 10$$

$$\therefore \qquad 33.3 \log (S_2/S_1) = 10$$

$$\therefore \qquad \log (S_2/S_1) = 0.3$$

$$\therefore \qquad S_2 = 2S_1$$

Thus an increase of 10 phons is the same as a doubling in the numerical value of the sones.

There are other ways in which loudness can be expressed. In fact different applications of acoustics have tended to develop, independently of each other, their own ways of describing what is basically the same information. Aircraft engineers frequently use **perceived noise level**, abbreviated PNL and measured in PNdB. It has been derived in the same statistical manner as phons, which is hardly surprising, but it is based on judgements of equal annoyance rather than of equal loudness. The difference may appear to be trivial, but it is possibly a better criterion to use when one is dealing with the annoyance due to relatively noisy aircraft movements.

There is one other difference in the definition; the PNL of a noise is defined as the SPL of noise covering the band 910 to 1090 Hz which sounds equally noisy. It should be used only when dealing with aircraft noise.

A unit of **noisiness**, called a **noy**, is related to PNL in the same way as sones are related to phons, i.e.

$$N = 2^{(PNL-40)/10}$$ (4.3)

If the PNL is increased by 10, then the noys value doubles.

Another method was developed by engineers who were concerned with noise interfering with communications in offices. They introduced the **speech interference level**, abbreviated SIL. The SIL is the arithmetic average of the sound pressure level in three octave bands, and the bands taken for this purpose are 500–1000 Hz, 1000–2000 Hz, and 2400–4000 Hz. This is approximately the range of frequencies covered by intelligible speech, and the average SPL over this range gives a rough indication by means of one number of how difficult it is likely to be for a person to make himself understood in a particular environment. The SIL applies to a spectrum covering a fairly large frequency range, it cannot be applied to a pure tone of a single frequency.

All these methods were developed to meet the requirements of particular problems, and have been found useful in their own fields. However, in all these cases, a single numerical value conveys no information about how the sound energy or sound pressure is distributed over the range of frequencies making up the total sound. If on the other hand we are given the SPLs in each of the eight octave bands, then we can calculate the total loudness level in accordance with any of the methods we wish.

To get an answer in phons the SPL of each octave band is first converted to phons from the equal loudness curves and then to sones according to eqn 4.2. There is an empirical formula for adding sones:

$$S_T = S_m + 0.3\,(\Sigma S - S_m)$$ (4.4)

where S_T is the total loudness, S_m is the loudness of the loudest band, and ΣS is the sum of all eight loudnesses. S_T is then converted back to phons to give the loudness level of the combined sound. We can thus predict the loudness of a complex sound by adding the loudnesses of the individual components. If we added the individual intensities, we would have the overall intensity, but this would not necessarily be any good guide to the subjective loudness.

Perceived noise level is calculated in a very similar way, by first converting the octave band SPLs to noys. The noys are then added according to the formula

$$N_T = N_m + 0.3\,(\Sigma N - N_m)$$

where N_T is the total noisiness, N_m is the largest single value of noisiness, and ΣN is the sum of all the eight octave band noys. Finally the noys are converted back to the PNL, using the counterpart of eqn 4.2.

$$PNL = 40 + 33.3 \log N_T$$ (4.5)

The methods of calculating loudness have been described for one sound covering a range of frequencies. If two or more sounds occur simultaneously, then before we can calculate the resulting loudness, we must combine the SPLs in each octave band, by the method described in Chapter 2.

4.3   Noise criteria

Another approach has been adopted to provide a standard of what is an acceptable level in given circumstances. Once again the opinion of a large number of people has been averaged, and the result expressed as a set of **noise criteria** or **NC curves**. These are curves of SPL against frequency as illustrated in fig. 4.3. The variation

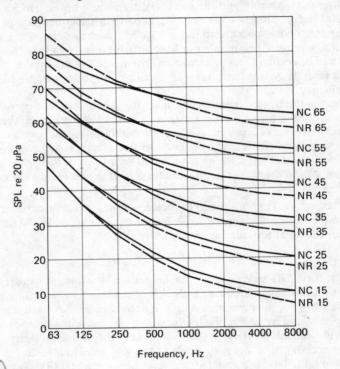

4.3 NC and NR curves

of SPL with frequency is more or less arbitrary, but the particular form shown was adopted because it was found that most people considered this particular distribution of sound energy to be less objectionable than other distributions. Each NC curve is determined by the eight octave band SPLs which together make up the total sound represented by the NC curve.

The curves have the property that the loudness level (LL) in phons is 22 units greater than the SIL. For example the curve NC 35 has an SIL of 35 dB and an LL of 57 phons. The curves also reduce in intensity with increasing frequency. The number which is arbitrarily given to the NC curve is its SIL.

Recommendations have been made for acceptable NC values in different environments, and the curves are generally used in conjunction with some such standard of acceptable values. A probable, anticipated, or measured, sound distribution is compared with the standard curves, and an NC number is assigned to the sound so as to correspond to the nearest NC curve which lies wholly above the sound distribution in question. The curves go up in steps of 5, and the nearest standard curve must be used; it is meaningless to interpolate between the standard curves. Having found an NC number for the sound in question, we can comment on whether it will be acceptable or not in any particular circumstances or environment. Alternatively we can specify that the sound level in a certain room must not exceed a particular NC value.

An alternative set of noise criteria, referred to as NC(A) curves, has also been published. These differ from the NC curves in that the difference between the LL and the SIL is 30 instead of 22. They are not recommended for normal use. The other set of curves shown in fig. 4.3 are **noise rating** or **NR curves**. These were designed, again on a statistical basis, for judging noise nuisance in private houses near industrial premises. For most practical purposes they may be regarded as identical to the NC curves.

4.4   Sound level meters

All methods of judging loudness are related in some way to sound pressure levels. The acoustic pressure can be detected and measured fairly readily. All sound level meters contain some device for converting the acoustic pressure into a directly proportional alternating electrical current or voltage. The value of the acoustic pressure can then be indicated in a number of ways. The most simple method is to read a single overall sound pressure level, but as we have seen this does not give a very good guide to the impression of loudness. The most detailed information is obtained by taking the electrical output of the sound level meter to an electrical frequency analyser. An electrical network can be designed to pass signals within any desired frequency range and to block signals of any other frequency. An acoustic analyser contains networks tuned to the different octave band frequencies, and separate readings of octave band SPLs are obtained by tuning the analyser to each band in turn. If a more detailed analysis is required it is possible to use an analyser which can be tuned to frequency bands each covering only one third of an octave. This information can be used directly, or it can be used to calculate sones, phons, perceived noise, or noys; or the data can be compared with the standard NC or NR curves.

It is not always convenient to use a frequency analyser because the analyser adds significantly to the size, weight and cost of the whole instrument. Instead, a frequency weighting network can be built into the sound level meter as shown in fig. 4.4 so that the output is no longer the total or overall sound pressure level. The sound pressure is always converted into a proportional electrical signal which is amplified and then measured. With a weighting network there is a different amount of amplification for each frequency so that the overall measurement finally made gives greater emphasis to some frequencies than to others. It will

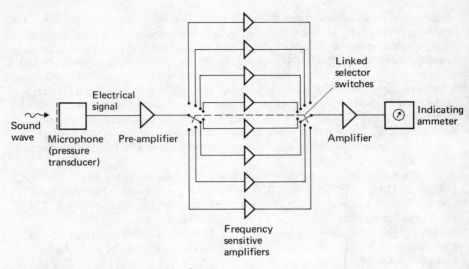

4.4 A block diagram of a sound level meter

be seen from the equal loudness contours that the ear is more responsive to
frequencies between 500 and 5000 Hz than to those above or below these frequencies. The weighting networks therefore allow greater amplification in this
range than at higher and lower frequencies. Three standard weighting networks have
been adopted, which are an attempt to approximate to the equal loudness curve.
The weightings are shown in fig. 4.5. The A scale is intended for use in measuring
sound pressures below 55 dB and the response approximates to the 40 phon equal
loudness contour. Measurements with this weighting are commonly expressed as
dB(A). The B scale is intended for use in the range 55 to 85 dB and the response
approximates to the 70 phon. equal loudness contour. The C scale, which is intended
for use above 85 dB, has an approximately flat response except at the two extremes
of the frequency range. Even with the use of these weighting networks a reading
of the sound level meter cannot be correlated with methods of judging loudness.

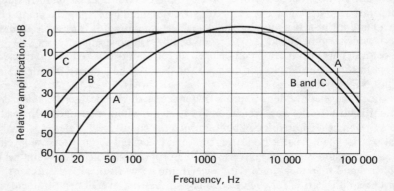

4.5 The frequency weightings

Example 43

For an accurate analysis the sound level meter is used with the C network and the output is fed into an octave band analyser. In spite of all the more sophisticated techniques for indicating loudness, the dB(A) scale has been found to be as good as any for giving a quick but accurate indication of how acceptable a given sound level will be.

It is in some ways a pity that so many different units should have been introduced into the study of sound; but this is typical of the piecemeal development of any subject where different workers are interested in different aspects of the same thing.

4.5 Example

As an example we will consider sound levels which were measured outside a block of flats. SPLs were measured in six octave bands, and were as follows:

Band mid-frequency Hz	63	125	250	500	1000	2000
SPL dB re 20 μPa	62	53	42	47	45	47

Comparison with the NC curves shows that at 2000 Hz the measured value is above NC 45 but below NC 50; so the sound must be described as being below NC 50. The fact that at 250 Hz it is below the NC 30 curve is irrelevant; the correct description for this overall sound is NC 50. The SIL is about 46 dB.

To find the overall level in phons we first convert each octave band SPL to phons from the equal loudness contours and then convert the phons to sones.

Band mid-frequency Hz	63	125	250	500	1000	2000
phons P	20	22	28	46	45	47
$(P-40)/10$	-2	-1.8	-1.2	0.6	0.5	0.7
$\log S = \dfrac{P-40}{10} \log 2$	-0.6	-0.54	-0.36	0.18	0.15	0.21
sones S	0.25	0.28	0.44	1.5	1.4	1.6

The loudest band is the 2000 Hz band, giving $S_m = 1.6$. The sum of all the loudnesses is $S = 2.47$ and the total loudness is

$$S_T = 1.6 + 0.3(2.47 - 1.6) = 1.6 + 0.3 \times 0.87$$

$$= 1.6 + 0.26$$

$$= 1.86 \text{ sones}$$

The total loudness in phons is then

$$P = 40 + 33.3 \log 1.86$$

$$= 40 + 33.3 \times 0.27 = 40 + 9$$

$$= 49 \text{ phons}$$

The overall SPL is found by taking the octave band pressures in pairs and adding the appropriate correction to the larger one. The results are again combined in pairs until the overall pressure is found.

Band mid-frequency Hz	63	125	250	500	1000	2000
SPL dB	62	53	42	47	45	47
correction		0.5		1		2
SPL dB		62.5		48		49
correction					2.5	
SPL dB		62.5		51.5		
correction			0.5			
SPL dB			63			

In this example we could, without significant error, have ignored all except the two largest octave band pressures. Here the low frequency pressure is the most significant one, whereas for both the NC rating and the overall loudness the high frequency was more significant. This happens only because of the way the energy is distributed over the frequencies of this particular sound. It would not necessarily be true of some other sound with some other pressure distribution. We cannot meaningfully convert the overall SPL to phons because it is not associated with any particular frequency.

To estimate what the A weighted reading of a sound level meter would be for this sound, we must apply the weighting factors to the octave band levels before combining them.

Band mid-frequency Hz	63	125	250	500	1000	2000
SPL dB	62	53	42	47	45	47
A weighting factor	−26	−16	−8	−4	0	+2
Weighted SPL dB(A)	36	37	34	43	45	49
correction		2.5		0.5		1.5
SPL dB(A)		39.5		43.5		50.5
correction					0.5	
SPL dB(A)		39.5			51	
correction			0			
SPL dB(A)			51			

It can be seen that the A weighting shifts the emphasis from the low frequencies to the higher ones, in a very similar way to the equal loudness curves, and for exactly the same reason, to conform more closely to the natural response of the human ear.

The calculation of noys would be done in the same way as that for phons.

4.6   Significance of measurements

These are some of the methods which various people have used to give an indication of the loudness of a sound. The present author has found that it is generally better to work in octave band levels and not to attempt to combine them in any way. The NC and NR curves are useful when it is required to predict whether a particular sound level, whose octave band pressure levels have been estimated, will be acceptable in certain circumstances or not. Their advantage, of course, lies in the fact that they give comparison values for each octave band. If one is dealing with an existing sound for which the pressure levels have actually been measured, then the human ear is the best judge of whether the sound is too loud or not. The measurements are merely the first, and not necessarily the most important, step in finding a way of reducing the sound if it is too loud. Because so many of the relevant factors vary with frequency, any attempt to combine octave band levels when investigating existing noises is likely to produce misleading conclusions.

It can be seen from the method of combining pressure levels of different frequencies that the overall sound pressure level must be greater than that of a single octave band (unless the sound is restricted to only one octave). In the same way, if SPLs are given at third-octave band intervals, the three levels which together make up one octave will be less than the overall pressure for that octave. Theoretically, SPLs could be plotted at intervals of 1 Hz, and these would be still lower.

There are two ways in which sound intensities can be shown on a graph. They can be drawn as a continuous curve of an intensity that varies continuously with frequency (fig. 4.6a). The ordinate in this case is intensity per unit frequency. The overall intensity in a bandwidth Δf is the area under the curve. The value of the overall intensity can be plotted as a single ordinate at the mid-point of each frequency band (fig. 4.6b). If the continuous spectrum is flat, then the effect of doubling the bandwidth chosen is to double the corresponding intensity, which means raising the intensity level by 3 dB.

It is in fact more convenient to work in terms of decibels, and as intensity level is very nearly equal to sound pressure level, the usual way of showing sound levels is to plot the octave band sound pressure level at the octave band mid-frequency, and then join the points by straight lines (fig. 4.6c). This corresponds with what is actually measured by a sound meter and analyser. In this case the line joining two points has no particular significance, and trying to read pressure levels in between the points is meaningless. If third-octave bands are used there will be three times as many points in a given frequency range, again joined by arbitrary straight lines which serve merely to join the points and convey no other information.

All these are measurements of sound pressure. They cannot on their own give any knowledge about the sound power output of a source. If, as is sometimes attempted, they are to be used to give an indication of how noisy a particular piece of machinery is likely to be, there must be some explicit or implied standard method of measuring the sound pressure under standard conditions. Frequently the measurements are taken in a free field at a distance of 1 metre from the machine. Since the machine is most unlikely to be installed in the identical conditions in which the measurements were made, they are of little value except as the first step in finding the machine's sound power level. Sometimes the measurements are made at a distance which depends on the size of the machine, such as three times the diameter of a fan. This

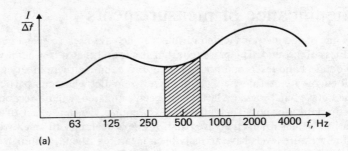

(a)

Shaded areas in
(a) and (b) are equal

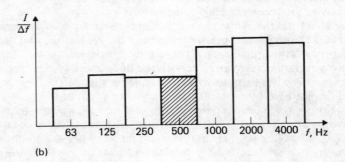

(b)

Ordinates in (c) are
proportional to areas in (b)

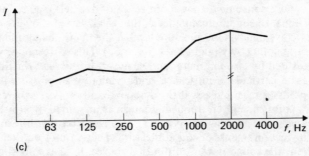

(c)

4.6 The graphical representation of intensity

may be convenient when doing the tests, but it makes it harder to compare the sound output of different size machines. Sometimes the quoted loudness level is what might be expected with the machine installed in a 'typical' room. When we come to look more closely at the many factors which affect the sound level in a room we will see just how misleading, if not completely devoid of meaning, such descriptions can turn out to be.

4.7   Pitch and frequency

We have discussed at some length the subjective quality of loudness and the various methods of correlating it with objective measurements of pressure and frequency. The other two subjective qualities are pitch and timbre.

Pitch is determined mainly by the frequency, but it also depends to some extent on the intensity and the wave form of the sound. For two sounds of the same frequency, the louder one has the lower pitch. If we wish to keep the pitch of a pure tone constant, then as the loudness increases it may be necessary to increase the frequency. This effect is particularly noticeable at low frequencies, and is at its greatest at 100 Hz. It is not very important between 1000 and 5000 Hz. The relationship between pitch and loudness is most obvious for a pure tone, that is for a pure sine wave.

All sounds encountered in practice, which are not deliberately produced for test purposes, are complex, with components of different frequencies and amplitudes. The pitch of such a sound will depend on the frequencies and amplitudes of all the components. For a complete series of harmonics, all having the same loudness, the pitch is determined by the smallest difference between the frequencies. The notes produced by a musical instrument contain a number of harmonics, the effect of which is to reduce the dependence of the instrument's pitch on loudness.

Some sounds have components whose frequencies are very close together. Such sounds have no recognizable pitch and can only be described as a noise. If the frequencies cover the whole of the audible range and the intensities are equal the noise is described as white noise, by analogy with white light which is made up of equal intensities of all the colours of the visible spectrum. White noise has a sizzling or rushing sound.

Just as the phon has been introduced to give some numerical measure of the subjective quality of loudness, so the **mel** has been used to attempt to measure the quality of pitch. A tone of 1000 Hz is said to have a pitch of 1000 mels. A tone which appears to have twice the pitch is then described as 2000 mels.

Two sounds which have very nearly the same frequency will be heard as having the same pitch. The amplitude, and so the loudness of the combined sound, however, varies with a frequency which is equal to the difference between the two basic frequencies. This variation in loudness is heard as a **beat**. If the beat frequency is adjusted to zero then the two basic frequencies are equal, and this provides a very precise method of adjusting two frequencies to be equal to each other.

If the difference between two frequencies is itself in the audible range it will be heard as a distinct difference tone rather than as a beat. If the output from the ear to the auditory nerve were directly proportional to the sound pressure input, two different frequencies would be heard as two different tones. But if the output contains a term which is not linearly proportional to the input, then sum and difference tones will appear in the output. Suppose for example that the response of the ear to a sound pressure p is given by

$$r = ap + bp^2$$

If two frequencies are present

$$p = p_1 \cos 2\pi f_1 t + p_2 \cos 2\pi f_2 t$$

Then the output is

$$r = a(p_1 \cos 2\pi f_1 t + p_2 \cos 2\pi f_2 t) + b(p_1 \cos 2\pi f_1 t + p_2 \cos 2\pi f_2 t)^2$$

Expanding the second term gives

$$r = a(p_1 \cos 2\pi f_1 t + p_2 \cos 2\pi f_2 t) + b(p_1^2 \cos^2 2\pi f_1 t$$

$$+ 2p_1 p_2 \cos 2\pi f_1 t \times \cos 2\pi f_2 t + p_2^2 \cos^2 2\pi f_2 t)$$

$$\therefore \quad r = a[p_1 \cos 2\pi f_1 t + p_2 \cos 2\pi f_2 t] + b[(p_1^2 + p_2^2)/2 + (p_1 \cos 4\pi f_1 t)/2$$

$$+ (p_2 \cos 4\pi f_2 t)/2 + p_1 p_2 \cos 2\pi (f_1 + f_2)t + p_1 p_2 \cos 2\pi (f_1 - f_2)t]$$

The output is thus the sum of seven components each of different frequency. Two of them are the original input frequencies, and one is a constant of zero frequency. Another two components contribute frequencies exactly double those of the input frequencies. Then there is one component with a frequency equal to sum of the input frequencies, and finally there is a component whose frequency is equal to the difference of the input frequencies. The existence of these difference tones is due to the non-linear response of the ear.

4.8   Timbre

Timbre is the quality which enables us to distinguish between the same note when played by different instruments. Thus middle C played on a piano sounds different from middle C played on a violin, even if they are equally loud, although the pitch is the same. The difference in timbre is partly due to the presence of different harmonics, which give the sounds different wave forms; but timbre also depends on both intensity and frequency. This can be shown by playing a recording at some loudness level, and then playing it again at a level some twenty phons louder. There will be a noticeable change in timbre. The effect of frequency can be similarly shown by playing a recording at the wrong speed. If the change in speed and therefore in frequency is enough, the timbre will be changed to such an extent that it becomes difficult to recognize the instruments. Although it has an obvious importance in music, considerations of timbre do not play an important part in the practical applications of acoustics.

4.9   Masking

When two sounds are heard simultaneously it may be difficult to distinguish between them. It is difficult, for example, to carry on a telephone conversation against a background of noisy traffic, or against the roar of overhead aircraft. We say that the louder sound **masks** the quieter. If one of two sounds is more than about twice as loud as the other, the quieter one will make a negligible contribution to the total loudness. If the difference in loudness is sufficiently great, then the quieter sound is inaudible in the presence of the louder. The formal definition states that aural masking is the increase in the threshold of hearing of the masked sound in the presence of the masking sound. It is measured in decibels relative to the threshold of hearing in the absence of the masking sound.

The masking effect is of some practical importance. It means that there is no point in trying to produce a machine or plant which is absolutely silent, because the sound from it will always be heard against a background of other sounds which to some extent will mask it. All that is required of a ventilation system, for example, is a sound level sufficiently low to be inaudible against the normal level of noise in the room being ventilated. Obviously the appropriate level will be lower in a concert hall than in a factory, and the normal ambient sound levels must be considered in deciding on the acoustic treatment required in any particular case.

In general, if an unwanted sound is to be kept to a level which will be unobjectionable, the SPL in each octave band should be 5 to 10 dB below that of the ambient sound. The intruding sound will then be barely noticeable. This is the application of the NC and NR curves; they provide a criterion by which we can judge what octave band pressure levels can be accepted for an unwanted sound intruding into any particular surroundings.

5 Transmission through walls

5.1 Sound paths

Sound can be transmitted from a source to a receiver in several ways, and along different paths. The various paths encountered in practice can be discussed in two very broad groups, airborne noise and structure-borne noise. Airborne noise can in turn be considered as falling into two closely related sub-groups. One of these is concerned with the transmission of noise through ducts, windows, and similar air passages. The other is given the rather odd description of 'transmission of airborne noise through the structure'; it is noise transmitted through the air but at some point in its path it passes through a partition forming part of a structure. Although the partition forms a relatively short part of the total path, its presence has a very great effect on the overall transmission.

We will discuss this kind of transmission first. It can be illustrated by thinking of two adjacent rooms, separated by a solid wall or partition, with a loudspeaker or other source of sound in one of them. We will refer to this room as the sending or source room, and to the other room as the receiving room. Let us imagine the actual source of sound suspended in mid-air such that there is no possibility of setting up vibrations in the walls, floor or ceiling by direct contact with them. Then all the sound radiated from it will be carried to the walls of the source room as airborne sound. When the sound reaches the partition between the source room and receiving room, some of it will be transmitted through the partition into the receiving room where it will appear as sound waves travelling through the air. The airborne sound has been transmitted through the structure.

Some of the energy reaching the partition is reflected back into the source room, some is transmitted into the source room, and a very small amount may be lost in the form of heat in the material of the partition.

5.2 Normal incidence at a boundary

We will investigate more closely what happens when sound waves reach a boundary between one medium and another. As always, the theory involves a number of simplifying assumptions, and since we have already seen that all waves approximate to plane waves except at points very close to the source, we will restrict our attention to plane waves. We will also assume that there is no loss of energy in either of the two media. Consider first a plane wave normally incident on a boundary. Referring to fig. 5.1 the equations for the incident, reflected, and transmitted pressure waves can be written:

$$p_i = A_1 \cos 2\pi (ft - x/\lambda_1) \tag{5.1}$$

$$p_r = B_1 \cos 2\pi (ft + x/\lambda_1) \tag{5.2}$$

$$p_t = A_2 \cos 2\pi (ft - x/\lambda_2) \tag{5.3}$$

The reflected wave travels in the opposite direction to the incident wave, and this is indicated by the change of sign in the bracket. The velocity of the incident wave is $f_1 \lambda_1 = c$ and that of the reflected waves is $f(-\lambda_1) = -c$.

5.1 The notation for reflection at a boundary

The wave length is different in different media, so we use λ_1 for wave length in the first medium and λ_2 for the wavelength in the second medium. The letter A, with the appropriate subscript to indicate the medium, will be used for the amplitude of waves travelling in one direction, and the letter B, again with the appropriate subscript, for the amplitude of waves travelling in the opposite direction. The corresponding particle velocities are, from eqn 2.5

$$u_i = (A_1/\rho_1 c_1) \cos 2\pi (ft - x/\lambda_1)$$
$$u_r = (-B_1/\rho_1 c_1) \cos 2\pi (ft + x/\lambda_1)$$
$$u_t = (A_2/\rho_2 c_2) \cos 2\pi (ft - x/\lambda_2)$$

The reflected wave is travelling in the negative direction, so the particle velocity will be negative for a positive pressure. This may be shown very readily by rewriting the derivation of eqn 2.5 with a positive sign in the bracket.

Let us take the boundary as being at $x = 0$. To preserve continuity, the pressures and the particle velocities normal to the boundary must be the same on both sides of the boundary. In other words, the sum of the incident and reflected pressures must at all times be equal to the transmitted pressure at $x = 0$, and in particular at time $t = 0$. The same requirement is true of the particle velocities. These conditions can be stated as

$$p_i + p_r = p_t, \text{ at } x = 0$$
$$u_i + u_r = u_t, \text{ at } x = 0$$

Writing these out in full at the boundary $x = 0$,

$$A_1 \cos 2\pi ft + B_1 \cos 2\pi ft = A_2 \cos 2\pi ft$$

$$(A_1/\rho_1 c_1) \cos 2\pi ft - (B_1/\rho_1 c_1) \cos 2\pi ft = (A_2/\rho_2 c_2) \cos 2\pi ft$$

Putting $t = 0$ gives

$$A_1 + B_1 = A_2 \tag{5.4}$$

$$(A_1 - B_1)/\rho_1 c_1 = A_2/\rho_2 c_2 \tag{5.5}$$

$$\therefore (A_1 - B_1)\rho_2 c_2/\rho_1 c_1 = A_2 = A_1 + B_1$$

$$\therefore B_1 \left(1 + \frac{\rho_2 c_2}{\rho_1 c_1}\right) = A_1 \left(\frac{\rho_2 c_2}{\rho_1 c_1} - 1\right)$$

$$\therefore B_1 = A_1 \frac{\rho_2 c_2 - \rho_1 c_1}{\rho_1 c_1} \frac{\rho_1 c_1}{\rho_1 c_1 + \rho_2 c_2}$$

$$\therefore \frac{B_1}{A_1} = \frac{\rho_2 c_2 - \rho_1 c_1}{\rho_2 c_2 + \rho_1 c_1} \tag{5.6}$$

Eqn 5.6 gives the ratio of the reflected pressure amplitude to the incident pressure amplitude. The intensity in any one medium is proportional to the square of the pressure, so the ratio of the intensities will be the square of the pressure ratio. Since the area of the wave front is the same for both the incident and reflected waves, the ratio of the powers is equal to the ratio of intensities. We can thus define a **sound power reflection coefficient** as

$$\alpha_r = \left(\frac{\rho_2 c_2 - \rho_1 c_1}{\rho_2 c_2 + \rho_1 c_1}\right)^2 \tag{5.7}$$

We can derive a sound power transmission coefficient in a similar way. From eqn 5.4

$$(A_1 + B_1)/\rho_1 c_1 = A_2/\rho_1 c_1$$

Adding eqn 5.5

$$\frac{2A_1}{\rho_1 c_1} = A_2 \left(\frac{1}{\rho_1 c_1} + \frac{1}{\rho_2 c_2}\right) = A_2 \left(\frac{\rho_2 c_2 + \rho_1 c_1}{\rho_1 c_1 \rho_2 c_2}\right)$$

$$\therefore \frac{A_2}{A_1} = 2 \frac{\rho_2 c_2}{\rho_2 c_2 + \rho_1 c_1}$$

Now the ratio of the intensities is

$$\frac{A_2^2}{\rho_2 c_2} \frac{\rho_1 c_1}{A_1^2} = \frac{\rho_1 c_1}{\rho_2 c_2} \left(\frac{2\rho_2 c_2}{\rho_2 c_2 + \rho_1 c_1}\right)^2$$

Since we have assumed plane waves, which do not diverge, the ratio of the energies is the same as the ratio of the intensities. So the **sound power transmission coefficient** is

$$\alpha_t = \frac{4\rho_2 c_2 \rho_1 c_1}{(\rho_2 c_2 + \rho_1 c_1)^2} \tag{5.8}$$

Eqns 5.7 and 5.8 are symmetrical in $\rho_1 c_1$ and $\rho_2 c_2$. This means that if the direction in which the incident wave is travelling is reversed, so that it reaches the boundary from medium 2, is reflected in medium 2 and transmitted into medium 1, the coefficients of reflection and transmission remain exactly the same as before.

It can also be seen that the sum of the two coefficients is 1, for

$$\alpha_r + \alpha_t = \frac{\rho_2^2 c_2^2 - 2\rho_2 c_2 \rho_1 c_1 + \rho_1^2 c_1^2}{(\rho_2 c_2 + \rho_1 c_1)^2} + \frac{4\rho_2 c_2 \rho_1 c_1}{(\rho_2 c_2 + \rho_1 c_1)^2}$$

$$= \frac{\rho_2^2 c_2^2 + 2\rho_2 c_2 \rho_1 c_1 + \rho_1^2 c_1^2}{(\rho_2 c_2 + \rho_1 c_1)^2} = \frac{(\rho_2 c_2 + \rho_1 c_1)^2}{(\rho_2 c_2 + \rho_1 c_1)^2}$$

$$= 1$$

This is what would be expected from the assumption that there is no loss of energy, for if no energy is lost then the whole of the incident energy is either reflected or transmitted, and the sum of the two fractions must be 1.

This theory applies fairly well to the boundary between air and a thick rigid solid, if the area of the boundary is large in relation to the wavelength. It does not apply to a thin or flexible solid. If the boundary is not sufficiently rigid it will vibrate as a membrane or thin plate, and the entire boundary will move in sympathy with the incident sound waves. Nor does the theory apply if one side of the boundary is formed by a porous solid. In a porous solid the fluid will flow into and out of the voids and some of the previous assumptions no longer hold. In particular, both the reflected and the transmitted waves are out of phase with the incident wave.

We assumed, without proof, that there was no change of phase when the pressure wave was reflected or transmitted at the boundary. That this was so can be shown by considering time $t = 1/4f$ and $x = 0$. Then if the reflected and transmitted waves differ in phase from the incident wave by ϕ_r and ϕ_t respectively, eqns 5.1 to 5.3 become

$$p_t = A_1 \cos(\pi/2) = 0$$

$$p_r = B_1 \cos(\pi/2 + \phi_r) = -B_1 \sin \phi_r$$

$$p_t = A_2 \cos(\pi/2 + \phi_t) = -A_2 \sin \phi_t$$

Proceeding as before, these give

$$-B_1 \sin \phi_r = -A_2 \sin \phi_t$$

and

$$\frac{B_1 \sin \phi_r}{\rho_1 c_1} = \frac{-A_2 \sin \phi_t}{\rho_2 c_2}$$

$$\therefore \quad \frac{\rho_2 c_2}{\rho_1 c_1} B_1 \sin \phi_r = -A_2 \sin \phi_t$$

$$= -B_1 \sin \phi_r$$

which is only possible if $\sin \phi_r = 0$ and therefore $\phi_r = 0$, in agreement with our assumption. We will then have, at $t = 1/4f$, $0 = A_2 \sin \phi_t$. This is again only possible if $\sin \phi_t = 0$ and $\phi_t = 0$, as assumed.

At a boundary between a fluid and a porous solid, conditions are not uniform over the wavefront, but vary between the solid and the pores, and the assumptions implicit in our analysis no longer hold. Phase changes occur and give rise to reflection and transmission coefficients which depend on frequency, thickness, porosity and density.

5.3   Normal transmission past two boundaries

Let us extend the analysis to the case of three media separated by two boundaries. The medium in the middle could, for example, be the wall of a house, or in a more general case, the side of a ship.

The incident wave at the first boundary is partly reflected and partly transmitted. When the transmitted wave reaches the second boundary it is in turn partly reflected and partly transmitted. The reflected part goes back to the first boundary where it is partly transmitted and partly reflected. The transmitted part adds to the first reflected wave. The reflected part returns to the second boundary and a process of repeated reflection occurs between the two boundaries. Ultimately a state of equilibrium will be reached which can be represented by the following equations for the pressure and particle velocities of the waves in each medium. In the first and second media there are two waves, travelling in opposite directions as shown in fig. 5.2.

$p_{i,1}$		$p_{t,2}$		$p_{t,3}$
$p_{r,1}$		$p_{r,2}$		
$u_{i,1}$ →		$u_{t,2}$ →		$u_{t,3}$ →
$u_{r,1}$ ←		$u_{r,2}$ ←		
A_1 →		A_2 →		A_3 →
B_1 ←		B_2 ←		

5.2 The notation for transmission past two boundaries

We will again apply the conditions that both pressure and particle velocity are continuous at both the boundaries, i.e. at $x = 0$ and $x = l$. The number of equations involved makes the maths tedious, but not difficult.
At $x = 0$ we have

$$p_{i,1} + p_{r,1} = p_{t,2} + p_{r,2}$$

$$u_{i,1} + u_{r,1} = u_{t,2} + u_{r,2}$$

At $x = l$ we have

$$p_{t,2} + p_{r,2} = p_{t,3}$$
$$u_{t,2} + u_{r,2} = u_{t,3}$$

Therefore

$$A_1 \cos 2\pi ft + B_1 \cos (2\pi ft + \phi_{B_1}) = A_2 \cos (2\pi ft + \phi_{A_2})$$
$$+ B_2 \cos (2\pi ft + \phi_{B_2})$$

$$\rho_2 c_2 A_1 \cos 2\pi ft - \rho_2 c_2 B_1 \cos (2\pi ft + \phi_{B_1}) = \rho_1 c_1 A_2 \cos (2\pi ft + \phi_{A_2})$$
$$+ \rho_1 c_1 B_2 \cos (2\pi ft + \phi_{B_2})$$

$$A_2 \cos (2\pi ft - 2\pi l/\lambda_2 + \phi_{A_2}) + B_2 \cos (2\pi ft + 2\pi l/\lambda_2 + \phi_{B_2})$$
$$= A_3 \cos (2\pi ft - 2\pi l/\lambda_3 + \phi_{A_3})$$

$$\rho_3 c_3 A_2 \cos (2\pi ft - 2\pi l/\lambda_2 + \phi_{A_2}) - \rho_3 c_3 B_2 \cos (2\pi ft + 2\pi l/\lambda_2 + \phi_{B_2})$$
$$= \rho_2 c_2 A_3 \cos (2\pi ft - 2\pi l/\lambda_3 + \phi_{A_3})$$

By putting $t = 0$ and $t = 1/4f$ we will obtain eight equations, from which we can get eight unknowns $A_2, A_3, B_1, B_2, \phi_{A_2}, \phi_{A_3}, \phi_{B_1}, \phi_{B_2}$ in terms of the known quantities $A_1, \rho_1 c_1, \rho_2 c_2, \rho_3 c_3$.

The solution of a set of eight simultaneous equations is tedious and the full working in terms of trigonometric functions becomes very long-winded. It is very much easier to re-write the equations in terms of complex exponential functions, since one complex number gives both amplitude and phase. Treating B_1, B_2, B_3, A_1, A_2 and A_3 as complex numbers we can use the notation and method used in Appendix 1 for the earlier mathematics.

We then have at $t = 0, x = 0$,

$$A_1 + B_1 = A_2 + B_2 \tag{5.9}$$

$$\rho_2 c_2 (A_1 - B_1) = \rho_1 c_1 (A_2 - B_2) \tag{5.10}$$

and at $t = 0, x = l$

$$A_2 \exp(-jk_2 l) + B_2 \exp(jk_2 l) = A_3 \exp(-jk_3 l) \tag{5.11}$$

$$\rho_3 c_3 [A_2 \exp(-jk_2 l) - B_2 \exp(jk_2 l)] = \rho_2 c_2 A_3 \exp(-jk_3 l) \tag{5.12}$$

From eqn 5.9

$$\rho_2 c_2 (A_1 + B_1) = \rho_2 c_2 (A_2 + B_2)$$

adding eqn 5.10

$$2\rho_2 c_2 A_1 = A_2 (\rho_2 c_2 + \rho_1 c_1) + B_2 (\rho_2 c_2 - \rho_1 c_1) \tag{5.13}$$

From eqn 5.11

$$\rho_3 c_3 [A_2 \exp(-jk_2 l + B_2 \exp(jk_2 l)] = \rho_3 c_3 A_3 \exp(-jk_3 l)$$

adding eqn 5.12

$$2\rho_3 c_3 A_2 \exp(-jk_2 l) = (\rho_3 c_3 + \rho_2 c_2) A_3 \exp(-jk_3 l) \tag{5.14}$$

Subtracting eqn 5.12

$$2\rho_3 c_3 B_2 \exp(jk_2 l) = (\rho_3 c_3 - \rho_2 c_2)A_3 \exp(-jk_3 l) \tag{5.15}$$

substituting eqns 5.14 and 5.15 in eqn 5.13

$$2\rho_2 c_2 \frac{A_1}{A_3} = \frac{(\rho_2 c_2 + \rho_1 c_1)(\rho_3 c_3 + \rho_2 c_2)}{2\rho_3 c_3} \frac{\exp(-jk_3 l)}{\exp(-jk_2 l)}$$

$$+ \frac{(\rho_2 c_2 - \rho_1 c_1)(\rho_3 c_3 - \rho_2 c_2)}{2\rho_3 c_3} \frac{\exp(-jk_3 l)}{\exp(jk_2 l)}$$

$$= \frac{(\rho_2 c_2 + \rho_1 c_1)(\rho_3 c_3 + \rho_2 c_2)}{2\rho_3 c_3} \exp jl(k_2 - k_3)$$

$$+ \frac{(\rho_2 c_2 - \rho_1 c_1)(\rho_3 c_3 - \rho_2 c_2)}{2\rho_3 c_3} \exp[jl(-k_2 - k_3)]$$

$$\therefore \frac{A_1}{A_3} = \frac{\begin{aligned}(\rho_2 c_2 + \rho_1 c_1)(\rho_3 c_3 + \rho_2 c_2) \exp[jl(k_2 - k_3)] \\ + (\rho_2 c_2 - \rho_1 c_1)(\rho_3 c_3 - \rho_2 c_2) \exp[jl(-k_2 - k_3)]\end{aligned}}{4\rho_3 c_3 \rho_2 c_2} \tag{5.16}$$

Using $\exp j\theta = \cos\theta + j\sin\theta$ on eqn 5.16 gives

$$4\rho_3 c_3 \rho_2 c_2 \frac{A_1}{A_3} = (\rho_2 c_2 + \rho_1 c_1)(\rho_3 c_3 + \rho_2 c_2)\cos l(k_2 - k_3) + (\rho_2 c_2 - \rho_1 c_1)$$
$$\times (\rho_3 c_3 - \rho_2 c_2)\cos l(k_2 + k_3) + j(\rho_2 c_2 + \rho_1 c_1)(\rho_3 c_3 + \rho_2 c_2)\sin l$$
$$\times (k_2 - k_3) - j(\rho_2 c_2 - \rho_1 c_1)(\rho_3 c_3 - \rho_2 c_2)\sin l(k_2 + k_3)$$

Write this for the time being as

$$P\cos l(k_2 - k_3) + Q\cos l(k_2 + k_3) + jP\sin l(k_2 - k_3) - jQ\sin l(k_2 + k_3)$$
$$= P\cos lk_2 \cos lk_3 + P\sin lk_2 \sin lk_3 + Q\cos lk_2 \cos lk_3 - Q\sin lk_2 \sin lk_3$$
$$+ jP\sin lk_2 \cos lk_3 - jP\cos lk_2 \sin lk_3 - jQ\sin lk_2 \cos lk_3 - jQ\cos lk_2 \sin lk_3$$
$$= (P + Q)\cos lk_2 \cos lk_3 + (P - Q)\sin lk_2 \sin lk_3 + j(P - Q)\sin lk_2 \cos lk_3 -$$
$$j(P + Q)\cos lk_2 \sin lk_3 \tag{5.17}$$

The square of the real part of eqn 5.17 is

$$(P + Q)^2 \cos^2 lk_2 \cos^2 lk_3 + 2(P + Q)(P - Q)\sin lk_2 \cos lk_2 \sin lk_3 \cos lk_3$$
$$+ (P - Q)^2 \sin^2 lk_2 \sin^2 lk_3$$

The square of the imaginary part is

$$(P - Q)^2 \sin^2 lk_2 \cos^2 lk_3 - 2(P + Q)(P - Q)\sin lk_2 \cos lk_2 \sin lk_3 \cos lk_3$$
$$+ (P + Q)^2 \cos^2 lk_2 \sin^2 lk_3$$

Adding the squares of the real and imaginary parts gives the square of the magnitude as

$$(P+Q)^2 \cos^2 lk_2 (\cos^2 lk_3 + \sin^2 lk_3) + (P-Q)^2 \sin^2 lk_2 (\sin^2 lk_3 + \cos^2 lk_3)$$
$$= (P+Q)^2 \cos^2 lk_2 + (P-Q)^2 \sin^2 lk_2 \quad (5.18)$$

Now

$$(P+Q) = (\rho_2 c_2 + \rho_1 c_1)(\rho_3 c_3 + \rho_1 c_1) + (\rho_2 c_2 - \rho_1 c_1)(\rho_3 c_3 - \rho_2 c_2)$$
$$= 2\rho_2 c_2 \rho_3 c_3 + 2\rho_1 c_1 \rho_2 c_2$$
$$\therefore \quad (P+Q)^2 \cos^2 lk_2 = 4\rho_2^2 c_2^2 (\rho_3 c_3 + \rho_1 c_1)^2 \cos^2 lk_2 \quad (5.19)$$

Also

$$(P-Q) = (\rho_2 c_2 + \rho_1 c_1)(\rho_3 c_3 + \rho_2 c_2) - (\rho_2 c_2 - \rho_1 c_1)(\rho_3 c_3 - \rho_2 c_2)$$
$$= 2\rho_2^2 c_2^2 + 2\rho_1 c_1 \rho_3 c_3$$
$$\therefore \quad (P-Q)^2 \sin^2 lk_2 = 4(\rho_2^2 c_2^2 + \rho_1 c_1 \rho_3 c_3)^2 \sin^2 lk_2 \quad (5.20)$$

Dividing eqns 5.19 and 5.20 by $16\rho_3^2 c_3^2 \rho_2^2 c_2^2$ and adding gives

$$\left(\frac{A_1}{A_3}\right)^2 = \frac{(\rho_3 c_3 + \rho_1 c_1)^2 \cos^2 lk_2}{4\rho_3^2 c_3^2} + \frac{(\rho_2 c_2 + \rho_1 c_1 \rho_3 c_3 / \rho_2 c_2)^2 \sin^2 lk_2}{4\rho_3^2 c_3^2} \quad (5.21)$$

The ratio of the intensities is

$$(A_3^2 / \rho_3 c_3)(\rho_1 c_1 / A_1^2)$$

Substituting for A_3^2 / A_1^2 from eqn 5.21 gives the sound power transmission coefficient

$$\alpha_t = \left(\frac{A_3}{A_1}\right)^2 \frac{\rho_1 c_1}{\rho_3 c_3}$$
$$= \frac{4\rho_3 c_3 \rho_1 c_1}{(\rho_3 c_3 + \rho_1 c_1)^2 \cos^2 lk_2 + (\rho_2 c_2 + \rho_1 c_1 \rho_3 c_3 / \rho_2 c_2)^2 \sin^2 lk_2} \quad (5.22)$$

This is a cumbersome expression which we will have to simplify if we wish to make any use of it for routine calculations. The first simplification is to restrict ourselves to cases where the fluid is the same on either side of a partition, which enables us to put $\rho_3 c_3 = \rho_1 c_1$. Then

$$\alpha_t = \frac{4\rho_1^2 c_1^2}{(2\rho_1 c_1)^2 \cos^2 lk_2 + (\rho_2 c_2 + \rho_1^2 c_1^2 / \rho_2 c_2)^2 \sin^2 lk_2}$$
$$= \frac{4}{4 \cos^2 lk_2 + (\rho_2 c_2 / \rho_1 c_1 + \rho_1 c_1 / \rho_2 c_2)^2 \sin^2 lk_2} \quad (5.23)$$

This equation applies to walls and partitions in air, but it does not apply for example for sound transmitted from the sea through the side of a ship, where on one side we have sea water for which $\rho c = 1\cdot54 \times 10^6$ rayls and on the other air for which $\rho c = 415$ rayls. It would however apply for sound transmitted from water to water through a steel partition.

We can make a further simplification by considering the kind of partitions met in ordinary building construction. Thus for concrete $\rho c = 8.1 \times 10^6$ rayls while for air at room temperature $\rho c = 415$ rayls. These values give $\rho_2 c_2 / \rho_1 c_1 = 2 \times 10^4$ and $\rho_1 c_1 / \rho_2 c_2 = 5 \times 10^{-5}$. We may therefore ignore $\rho_1 c_1 / \rho_2 c_2$ in comparison with $\rho_2 c_2 / \rho_1 c_1$ and the expression reduces to

$$\alpha_t = \frac{4}{4 \cos^2 (2\pi l / \lambda_2) + (\rho_2 c_2 / \rho_1 c_1)^2 \sin^2 (2\pi l / \lambda_2)} \qquad (5.24)$$

The maximum value of $4 \cos^2 l / \lambda_2$ is 4. We can ignore this providing the other term in the denominator is greater than say 160. This requires approximately

$$\frac{\rho_2 c_2}{\rho_1 c_1} \sin (2\pi l / \lambda_2) > 12.7$$

$$\therefore \quad \sin (2\pi l / \lambda_2) > \frac{12.7}{2 \times 10^4}$$

$$> 6.35 \times 10^{-4}$$

$$\therefore \quad 2\pi l / \lambda_2 > 6.35 \times 10^{-4}$$

For a wall 0.1 m thick, this gives

$$1 / \lambda_2 > 6.35 \times 10^{-4} / 2\pi \times 0.1$$

$$\text{or} \quad \lambda_2 < 0.99 \times 10^5 \text{ metres}$$

Since in concrete $c = 3.1 \times 10^3 \text{ m s}^{-1}$, this corresponds to a frequency $f > 3.1 \times 10^{-2}$ Hz, i.e. for all frequencies in the audible range the $\cos^2$ term can be ignored.

If the frequency is sufficiently low, the wavelength will be large, and $\sin 2\pi l / \lambda_2$ will be small enough to write $\sin 2\pi l / \lambda_2 = 2\pi l / \lambda_2$. This approximation holds providing $2\pi l / \lambda_2$ is less than about 0.6, which makes λ_2 more than about 1 metre if l is 0.1 m. λ_2 is the wavelength in concrete, and the corresponding frequency is 3000 Hz, which is well into the seventh octave band. The approximation holds for all lower frequencies.

Making these two approximations in eqn 5.24 we have the theoretical result that for all practical solid partitions in buildings, except at very high frequencies, the coefficient of transmission is

$$\alpha_t = \frac{4\rho_1^2 c_1^2 \lambda_2^2}{4\pi^2 \rho_2^2 c_2^2 l^2} \qquad (5.25)$$

Even this can be reduced to a more useful form, by inserting $\rho_1 c_1 = 415$ rayls for air, since in practice we are not likely to be dealing with any other fluid. At the same time, since the frequency remains constant where c and λ do not, it is better to replace c_2 / λ_2 by f. Making these two changes gives

$$\alpha_t = (415/\pi)^2 / \rho_2^2 l^2 f^2$$

But $\rho_2 l$ is the mass of a unit area of the wall; if we use σ for this quantity, then

$$\alpha_t = 1.75 \times 10^4/\sigma^2 f^2 \tag{5.26}$$

5.4   Sound reduction index

For most calculation purposes it is more convenient to use the **sound reduction factor** or **sound reduction index**, SRI, which is defined as

$$SRI = 10 \log (1/\alpha_t)$$

Substituting eqn 5.26 gives

$$SRI = -10 \log (1.75 \times 10^4) + 10 \log (\sigma^2 f^2)$$
$$= -42.4 + 20 \log (\sigma f) \tag{5.27}$$

where σ is in kg m^{-2}.

It is important to appreciate the many assumptions that have been made in arriving at this entirely theoretical formula. First of all we have assumed that the sound is propagated as plane waves both in the partition and in the air on either side. We have assumed that the partition is a rigid solid and that no other vibrations are set up in it. This includes the assumption that the wall does not have any natural frequency of vibration which would give rise to resonance effects. We have assumed that there is no loss of acoustic energy by conversion to other forms of energy. We have made an approximation which is not valid at high frequencies. But above all, we have assumed that the incident wave is normal to the partition. In practice this is never so, for sound waves will reach a wall or partition from all directions. The effect is that the numerical coefficients in eqn 5.27 must be altered, but its general form does not change. The empirical equation is

$$SRI = -17 + 15 \log (\sigma f) \tag{5.28}$$

Not only is this less than the theoretical value, but it does not increase with weight and frequency as rapidly as theory predicts.

The sound transmission coefficient is sometimes defined in terms of energy rather than power or intensity. The sound reduction index is also called the **transmission loss**, and is sometimes used to refer to a reduction in sound power and sometimes to a reduction in sound energy. So long as we are dealing with normally incident plane waves there is no difference between any of these definitions, but in other cases it is as well to be clear which definition is being used.

The sound reduction index is a measure of the sound insulating value of the partition. It can be seen that it is higher for high frequencies and lower for low frequencies, and increases at about 4.5 or 5 dB per octave. One result of this frequency dependence is that both speech and music are distorted in passing through a partition. Whatever ones choice of music, it cannot be heard at its best through a wall which attenuates the high notes more than the low. Equally, one can often hear that people are talking on the other side of a wall but cannot distinguish what they are saying because the normal distribution of amplitude with frequency, that we are accustomed to in ordinary speech, is disturbed too much.

The sound reduction index also increases as the mass per unit area of the wall is increased. A doubling of mass increases the SRI by 6 dB; this is known as the **Mass Law**. In fact for single leaf constructions the mass of the wall is the most important factor in determining its sound insulating value. Changes of material will have a negligible effect unless they are accompanied by change of density. The actual material used for a wall can be chosen for reasons quite unconnected with acoustics; all the acoustic engineer is concerned with is mass per unit area, so that acoustic considerations will dictate at most the thickness of the wall. It is often possible to achieve useful results by lining a partition with a thin layer of a dense material such as lead.

5.5   Double walls

The behaviour of double leaf partitions is very much more complicated to analyse. There are four boundaries instead of the two with a single leaf partition, and although it is easy enough to write down the boundary conditions in exactly the same way as for the single leaf case, there will be twice as many simultaneous equations to solve. However, for any specific partition of known construction, all the quantities except the amplitudes can be replaced by numbers, and the equations can then be solved to give the numerical ratio of the transmitted to the incident amplitude for each frequency. If even this is felt to be too much like hard work the whole thing can be done on a computer one frequency at a time to give a series of numerical values. In fact this is the kind of problem which is ideally suited for an analogue computer.

The response is much more frequency dependent than that of single leaf constructions. We saw that for concrete the wavelength at 3000 Hz is about 1 metre, from which we may conclude that all normal walls are thin compared with the wavelength, except at very high frequencies. If at any time the displacement on one side is $A \sin 2\pi ft$, then the displacement on the other side at the same time is $A \sin 2\pi(ft + l/\lambda)$. If l/λ is small, we have just said that in almost all practical cases it is, then the two displacements are approximately equal, and the entire wall vibrates as one. At low frequencies, and therefore long wavelengths, the presence of an air gap between two leaves of a wall has little effect and the wall still vibrates as one. The stiffness of the air space is such that both leaves move together. Another way of putting it is to say that the wave is so long relative to the air space that the wave does not know the space is there. At higher frequencies the two leaves may be regarded as two independent partitions each of which follows eqn 5.28, and the total attenuation is the sum of the separate attenuations. (Under these conditions for a given total weight the attenuation will be roughly twice as great at a given frequency.)

One consequence is that to get the fullest possible benefit from double glazing it is essential to have a very large air gap between the two panes. Without a substantial air gap the extra insulation is achieved simply by increasing the mass of the window. In most cases double glazing is the most practical way of getting the extra mass needed for extra insulation, but we are fooling ourselves if we think that an air gap of a few centimetres is itself an attenuator. At 1000 Hz the wavelength in air is about a third of a metre, so that an air gap of less than about 0.3 m is not likely to add any attenuation to that due to the mass of the two panes of glass.

We have assumed that the partition does not absorb any of the energy reaching it, but that it transmits some and reflects the rest. In other words, it insulates by reflecting the sound energy back where it came from. With a double leaf partition it is perfectly feasible to fill the gap between the leaves with a sound absorbing material which will actually dissipate sound energy in the form of heat, so that the total of the reflected and transmitted energies is less than the incident energy.

To illustrate how ineffective this is, let us imagine a partition which reflects none of the incident energy but absorbs 95% of it and transmits the remaining 5%. The transmission coefficient is 0.05 and the sound reduction index is 10 log 20 = 10 x 1.3 = 13 dB. This should be compared with a sound reduction index of about 50 dB for a 0.3 m brick wall. Obviously, adding sound absorbent material to a wall will reduce the amount of sound passing through it, but the effect is small compared with the reduction due simply to the mass of the wall. Insulation is the property we are interested in when dealing with the transmission of sound from one space to another. Absorption is the property of interest when we are concerned with sound levels within one space or with the attenuation of a wave as it passes through one material. In general good absorbers are not the right materials to use where insulation is required.

5.6 Practical application

There are some practical details which have to be considered very carefully if the results achieved in practice are to bear any resemblance at all to the simple theoretical results. The theory requires us to imagine an infinitely long and infinitely high partition, for it has ignored completely anything that may happen at the edges of the partition. In all buildings sound can also travel along the longer path formed by the other walls, floors and ceilings which flank the wall being considered. This flanking transmission will reduce the actual amount of insulation obtained in practice and imposes a limit on what can be achieved with normal building construction. There is no point in improving the insulation of a wall if the flanking path transmits almost as much energy as the wall itself. In the same way, the insulation between the inside and outside of a building will not be improved by making the walls thicker if most of the sound in fact goes through the windows. The practical construction of double glazed windows may also allow so much energy to go round the edges of the panes, that the product when installed in a building does not give the same insulation as it did under very carefully controlled laboratory tests.

By assuming that the partition is homogeneous and continuous we have in fact ignored all discontinuities. For example, if we are considering a door, we have ignored the air gaps around it as well as any cracks or openings in it such as a letterbox. These can be thought of as special cases of flanking transmission, but it is a little easier to estimate their effect on the overall sound transmission. An overall sound transmission coefficient can be defined as

$$\alpha_t = \frac{S_1 \alpha_{t,1} + S_2 \alpha_{t,2}}{S_1 + S_2} \qquad (5.29)$$

where S_1 and S_2 are the areas of the two materials and $\alpha_{t,1}$ and $\alpha_{t,2}$ are the corresponding transmission coefficients. So long as we are talking of a partition which is thin in relation to the wavelength of the sound, we may deal with an air gap as if it were a material partition. Let the suffix 2 refer to a small area of poor insulation and so of high transmission coefficient. If, as in the case of a crack round a door, S_2 is less than 10% of S_1 and $\alpha_{t,2}$ is much greater than $\alpha_{t,1}$, then the denominator of eqn 5.29 is approximately S_1 and the numerator is approximately $S_2\alpha_{t,2}$ so,

$$\alpha_t \approx S_2\alpha_{t,2}/S_1$$

and

$$\mathrm{SRI} \approx 10 \log (S_1/S_2\alpha_{t,2})$$

For an air gap $\alpha_{t,2} = 1$ and $\mathrm{SRI} \approx 10 \log (S_1/S_2)$.

The inherent sound insulating properties of the door are not being used at all. All that has happened is that the sound energy transmitted has been reduced in the proportion by which the area of the total opening has been reduced by the material of the door. The overall sound transmission would be almost exactly the same whatever the sound reduction index of the door itself, because almost the whole of the sound energy which is transmitted goes through the crack. The same analysis applies to other examples such as windows reducing the insulating value of a brick wall.

5.7   Oblique incidence

It is worthwhile looking at the case of oblique incidence of plane waves at a boundary between two fluid media, for this will show that the ratio of the intensities is not always the same as the ratio of the powers. Generally it is power transmission that we are interested in, and so transmission coefficients usually refer to power rather than to intensity, but when using published data it is as well to find out on which definition they are based and exactly what it is that is being quoted.

So far, when we have considered the equations for a plane wave, it has been convenient to choose the x-axis in the direction of propagation of the wave. Now it is convenient to choose the x-axis in the direction normal to the boundary between the two media, and to use some other symbol, say d, for the direction in which the wave travels, see fig. 5.3. If the angle of incidence is θ, then distance in the direction of travel is given by

$$d = x \cos \theta + y \sin \theta$$

Distance in the negative direction of travel is given by

$$-d = -x \cos \theta + y \sin \theta$$

Sound waves obey the same laws of reflection and refraction as other waves; in particular the angle of reflection is equal to the angle of incidence, and the angle of refraction is related to the angle of incidence by Snell's Law

$$\sin \theta_i/\sin \theta_t = c_1/c_2$$

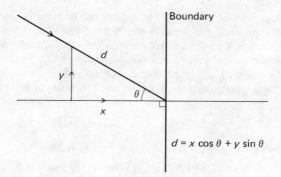

$$d = x \cos \theta + y \sin \theta$$

5.3 Oblique incidence

At any given frequency

$$c_1/c_2 = \lambda_1/\lambda_2$$

$$\sin \theta_t = (c_2/c_1) \sin \theta_i = (\lambda_2/\lambda_1) \sin \theta_i$$

$$\therefore \quad \sin \theta_t/\lambda_2 = \sin \theta_i/\lambda_1 \qquad (5.30)$$

Again using the complex exponential notation to avoid cumbersome trigonometric expressions, the equation for a wave travelling in the d-direction is

$$p = A \exp [j2\pi(ft - d/\lambda)]$$

$$= A \exp \{j2\pi [ft - (x/\lambda) \cos \theta - (y/\lambda) \sin \theta] \}$$

The incident, reflected, and transmitted waves, which were given by eqns 5.1 to 5.3 for the case of normal incidence, now become

$$p_i = A_1 \exp \{j2\pi [ft - (x/\lambda_1) \cos \theta_i - (y/\lambda_1) \sin \theta_i] \} \qquad (5.31)$$

$$p_r = B_1 \exp \{j2\pi [ft + (x/\lambda_1) \cos \theta_i - (y/\lambda_1) \sin \theta_i] \} \qquad (5.32)$$

$$p_t = A_2 \exp \{j2\pi [ft - (x/\lambda_2) \cos \theta_t - (y/\lambda_2) \sin \theta_t] \} \qquad (5.33)$$

At the boundary we have as before

$$p_i + p_r = p_t$$

The boundary is given by $x = 0$, but y can take any value. The boundary condition for pressure at time $t = 0$ therefore becomes

$$A_1 \exp [-(j2\pi y/\lambda_1) \sin \theta_i] + B_1 \exp [-j2\pi y/\lambda_1) \sin \theta_i] = A_2 \exp [-j2\pi y/\lambda_2) \sin \theta_t]$$

Substituting for $(\sin \theta_t/\lambda_2)$ from eqn 5.30 gives

$$A_1 \exp (-j2\pi y \sin \theta_i/\lambda_1) + B_1 \exp (-j2\pi y \sin \theta_i/\lambda_1) = A_2 \exp (-j2\pi y \sin \theta_i/\lambda_1)$$

$$\therefore \quad A_1 + B_1 = A_2 \qquad (5.34)$$

It is no coincidence that this is the same result as for the simpler case of normal incidence, for pressure at any point is the same in all directions. It can be described as a scalar quantity which does not depend on direction, so that changing the direction of the incident wave may change the appearance of some of the equations but will not change the physical relationship which they describe.

Velocity, however, is a vector quantity which has direction as well as magnitude, so we will expect the second boundary condition to lead to a result which does depend on the direction of incidence. The condition is that the components of velocity normal to the boundary are equal on the two sides of the boundary. This is written

$$u_i \cos \theta_i + u_r \cos (\pi - \theta_i) = u_t \cos \theta_t$$

$$\therefore \ (A_1/\rho_1 c_1) \cos \theta_i \exp \{j2\pi [ft - (x/\lambda_1) \cos \theta_i - (y/\lambda_1) \sin \theta_i] \}$$

$$-(B_1/\rho_1 c_1) \cos \theta_t \exp \{j2\pi [ft - (x/\lambda_2) \cos \theta_t - (y/\lambda_1) \sin \theta_i] \}$$

$$= (A_2/\rho_2 c_2) \cos \theta_t \exp \{j2\pi [ft - (x/\lambda_2) \cos \theta_t - (y/\lambda_1) \sin \theta_i] \}$$

Putting $x = 0$, $t = 0$, and cancelling the term $\exp(j2\pi y \sin \theta_i/\lambda_1)$, we have

$$\frac{A_1}{\rho_1 c_1} \cos \theta_i - \frac{B_1}{\rho_1 c_1} \cos \theta_i = \frac{A_2}{\rho_2 c_2} \cos \theta_t \tag{5.36}$$

As before, we eliminate A_2 between eqns 5.35 and 5.36, by substituting from eqn 5.35 in eqn 5.36 to give

$$A_1 \frac{\cos \theta_i}{\rho_1 c_1} - B_1 \frac{\cos \theta_i}{\rho_1 c_1} = A_1 \frac{\cos \theta_t}{\rho_2 c_2} + B_1 \frac{\cos \theta_t}{\rho_2 c_2}$$

$$\therefore \ A_1 \left(\frac{\cos \theta_i}{\rho_1 c_1} - \frac{\cos \theta_t}{\rho_2 c_2} \right) = B_1 \left(\frac{\cos \theta_i}{\rho_1 c_1} + \frac{\cos \theta_t}{\rho_2 c_2} \right)$$

$$\therefore \ \frac{B_1}{A_1} = \frac{\rho_2 c_2 \cos \theta_i - \rho_1 c_1 \cos \theta_t}{\rho_2 c_2 \cos \theta_i + \rho_1 c_1 \cos \theta_t} \tag{5.37}$$

The sound power reflection coefficient is given by

$$\alpha_r = \left(\frac{B_1}{A_1} \right)^2 = \left(\frac{\rho_2 c_2 \cos \theta_i - \rho_1 c_1 \cos \theta_t}{\rho_2 c_2 \cos \theta_i + \rho_1 c_1 \cos \theta_t} \right)^2 \tag{5.38}$$

Since all the incident power which is not reflected must be transmitted, the sound power transmission coefficient is

$$\alpha_t = 1 - \alpha_r$$

$$= \frac{(\rho_2 c_2 \cos \theta_i + \rho_1 c_1 \cos \theta_t)^2}{(\rho_2 c_2 \cos \theta_i + \rho_1 c_1 \cos \theta_t)^2} - \frac{(\rho_2 c_2 \cos \theta_i - \rho_1 c_1 \cos \theta_t)^2}{(\rho_2 c_2 \cos \theta_i + \rho_1 c_1 \cos \theta_t)^2}$$

$$= \frac{4 \rho_1 c_1 \rho_2 c_2 \cos \theta_i \cos \theta_t}{(\rho_2 c_2 \cos \theta_i + \rho_1 c_1 \cos \theta_t)^2} \tag{5.39}$$

The intensity of the transmitted wave can be found by eliminating B_1 between the two boundary conditions in eqns 5.35 and 5.36 giving

$$\frac{A_1}{\rho_1 c_1} \cos \theta_i - \frac{(A_2 - A_1)}{\rho_1 c_1} \cos \theta_i = \frac{A_2}{\rho_2 c_2} \cos \theta_t$$

$$\therefore \quad A_1 \left(\frac{2 \cos \theta_i}{\rho_1 c_1} \right) = A_2 \left(\frac{\cos \theta_i}{\rho_1 c_1} + \frac{\cos \theta_t}{\rho_2 c_2} \right)$$

$$\therefore \quad A_1 \, 2 \rho_2 c_2 \cos \theta_i = A_2 \, (\rho_2 c_2 \cos \theta_i + \rho_1 c_1 \cos \theta_t)$$

Then the coefficient of **sound intensity transmission** is

$$\alpha_i = \left(\frac{A_2^2}{\rho_2 c_2} \right) \left(\frac{\rho_1 c_1}{A_1^2} \right) = \left(\frac{A_2}{A_1} \right)^2 \left(\frac{\rho_1 c_1}{\rho_2 c_2} \right)$$

$$= \left[\frac{4 \rho_2^2 c_2^2 \cos^2 \theta_i}{(\rho_2 c_2 \cos \theta_i + \rho_1 c_1 \cos \theta_t)^2} \right] \left[\frac{\rho_1 c_1}{\rho_2 c_2} \right]$$

$$= \frac{4 \rho_1 c_1 \rho_2 c_2 \cos^2 \theta_i}{(\rho_2 c_2 \cos \theta_i + \rho_1 c_1 \cos \theta_t)^2} \tag{5.40}$$

Comparing eqns 5.39 and 5.40 we see that this time the value of the transmission coefficient depends on whether it is defined in terms of sound power or sound intensity. If the velocity in the second medium is greater than in the first, then from eqn 5.30

$$\sin \theta_i < \sin \theta_t$$

$$\therefore \quad \cos \theta_i > \cos \theta_t$$

$$\therefore \quad \cos^2 \theta_i > \cos \theta_i \cos \theta_t$$

and the intensity transmission coefficient is greater than the power transmission coefficient. The physical explanation of the difference is that the area of the wave front has changed.

From fig. 5.4 it can be seen that $AC = AB/\cos \theta_i = CD/\cos \theta_t$. So the areas of the

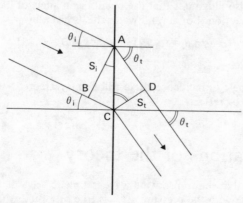

5.4 The change of wavefront area due to refraction

incident and refracted wavefronts, S_i and S_t respectively, are related by the following equation

$$S_i/\cos \theta_i = S_t/\cos \theta_t \tag{5.41}$$

If the powers are P_i and P_t and the corresponding intensities I_i and I_t, then by definition

$$I_i = P_i/S_i$$

and

$$I_t = P_t/S_t$$

$$\therefore \quad I_t/I_i = (P_t/P_i)(S_i/S_t)$$

$$= \alpha_t (S_i/S_t)$$

Substituting for α_t from eqn 5.39 and for S_i/S_t from eqn 5.41,

$$\alpha_i = \frac{I_t}{I_i} = \frac{4\rho_1 c_1 \rho_2 c_2 \cos \theta_i \cos \theta_t}{(\rho_2 c_2 \cos \theta_i + \rho_1 c_1 \cos \theta_t)^2} \frac{\cos \theta_i}{\cos \theta_t}$$

$$= \frac{4\rho_1 c_1 \rho_2 c_2 \cos^2 \theta_i}{(\rho_2 c_2 \cos \theta_i + \rho_1 c_1 \cos \theta_t)^2} \tag{5.40}$$

We thus see that the two coefficients must have different values because of the geometry of the refraction process.

Let us see how big the difference between the coefficients could be in practice. We will consider air for which $c = 343$ m s^{-1} and water for which it is 1481 m s^{-1}. Then $\sin \theta_t/\sin \theta_i$ is 4.3. If θ_i is $13°$, $\sin \theta_i$ is 0.225 and $\sin \theta_t$ is 0.974, making $\theta_t = 77°$. Then $\cos \theta_i$ is 0.974 and $\cos \theta_t$ is 0.225. The ratio of the two coefficients is

$$\alpha_i/\alpha_t = \cos \theta_i/\cos \theta_t = 0.974/0.225 = 4.33$$

So we see that it could be of some importance to know which transmission coefficient is being quoted.

The example also illustrates that there will be an angle of incidence which gives a $90°$ angle of refraction. For air and water this is given by

$$\sin \theta_i = (c_1/c_2) \sin \theta_t = 1/4.3 = 0.231$$

$$\therefore \quad \theta_i = 13° \, 22'$$

For any greater angle of incidence there will be no refracted wave and the whole of the energy will be reflected.

5.8   Limitations of the theory

We have assumed that the transmitted wave is a plane wave in which the particle velocity is in phase with the pressure. This is true of a fluid which cannot distort in shear or torsion so that the only waves present are longitudinal waves of the type we have been considering. A solid, on the other hand, can transmit other types of wave. However, the results we have obtained do hold for large, thick, rigid solids, such as most walls. They do not hold for very thin or flexible partitions such as sheet metal; material of this sort will vibrate as a whole, like a membrane, or like the skin of a drum. Nor do our results hold for porous materials. Here there is not only a compression wave in the solid, but there is also movement of the fluid in the pores. The pores are small (if they were not we would not be talking about a porous solid but about solid lumps floating in a fluid) and the fluid moves bodily in and out of the pores. Both these cases can be analysed by using more complicated equations to relate particle velocity to acoustic pressure, but in many cases it is quicker and easier to use experimentally measured values of reflection and transmission coefficients, which usually depend on frequency.

The problem becomes even more complicated when we consider oblique incidence. In some solids having a honeycomb structure, sound travels much more rapidly at right angles to the surface than parallel to the surface. In other solids the velocity of sound is very much smaller than in air, with the result that the angle of refraction is very much smaller than the angle of incidence. In both these types of solid the transmitted wave will travel in a direction very nearly normal to the surface boundary whatever the angle of incidence. The component of particle velocity normal to the surface will then have an almost constant ratio to acoustic pressure for all angles of incidence, and the theory outlined above must be modified accordingly. Many acoustic tiles and other sound absorbing materials behave in this way. The result is that the sound power reflection and sound power transmission coefficients vary with the angle of incidence in a more elaborate manner than the simple theory suggests.

For many solids the longitudinal compression wave is more important than any other type of wave and for these eqn 5.40 can be used to draw a graph of how the transmission coefficient varies with the angle of incidence. A more complicated analysis is required if the solid can transmit shear waves of significant magnitude. The possibilities are so varied and so involved that a more general analysis is not worthwhile. Except in advanced research work it is better to use empirical results.

Another way of analysing the transmission of sound through a wall considers that the sound waves which arrive at one side of the wall make the wall bend at the frequency of the sound. The whole wall bends, and the movement of the far side of the wall radiates sound into the air on that side. These ideas are developed a little more fully in Chapter 12. The mathematics is more difficult, but it leads to results almost exactly the same as those obtained in this chapter. There is however an extra result, which predicts that certain combinations of frequency and angle of incidence produce very high transmissions of acoustic energy through the walls. This is known as the **coincidence effect**, because the incident sound wave coincides with a natural mode of vibration of the wall.

6   Transmission through ducts

6.1   General considerations

There are many cases of sound being transmitted through ducting. In any ventilation system the fan is a potential source of noise which can travel along the ventilation ducts, but there are many other sources of noise that can be propagated through ventilation system ducting. Traffic noise can enter from outside a building, and other external sources may have to be considered in special cases. One important case is aircraft noise which must be taken into account in the vicinity of any airport. Within a building sound can be transmitted from one room to another through the ducting, which can be particularly embarrassing if one of the rooms is a private interview room. Ducts for the transmission of sound are also formed by gun barrels and motor car exhausts. Transmission of sound through ducts lends itself particularly easily to calculation, and for this reason the existence of other noise problems in a building with ventilation or air conditioning sometimes gets overlooked.

Whenever there is a sudden change in the cross-sectional area of the duct, some of the sound power is reflected back up the duct and there is a reduction in the sound power transmitted along the duct. A similar end reflection reduction occurs when a duct opens abruptly into a large cavity, with or without absorbent walls. This condition arises in practice when a ventilating duct terminates with a grille or diffuser in a room. The end reflection loss varies with the size of the duct and with the frequency of the sound. When a duct divides into two or more branches, in addition to any reflection, the transmitted power divides amongst the branches according to their areas, so that although there is no loss in total sound power, there is an effective attenuation in the amount transmitted along any open branch.

The main approximations which we shall make are that there is no reflection from the end of the pipe or from any other change of section except the one we are considering, and that there is no distortion of the wave front at the change of section.

6.2   Change of area

If the duct is large compared with the wavelength, then at any part of the wave front the sides are so far away that we may ignore them; the wave does not see the sides of the duct and continues on its way as if they were not there. In this case the fraction of the wave which is incident on the opening is transmitted without loss or reflection, and the fraction incident on the flange is completely reflected. Any loss, which in fact occurs at the flange, can be dealt with as part of the loss occurring at the rest of the duct surface. If S_1 is the area of a large duct from which sound is travelling into a smaller duct of area S_2 then the fraction of the sound power transmitted is

$$\alpha_t = S_2/S_1$$

and the fraction reflected is

$$\alpha_r = (S_1 - S_2)/S_1$$

When the sound passes from a small duct into a large one, or into a large space or cavity, almost the whole of the sound energy is transmitted so long as the opening is large in relation to the wavelength. This is what happens at high frequencies. At low frequencies the wavelength is larger than the duct dimensions and the sides of the duct play a part in what happens.

Approximate equations to give some idea of the amount of reflection occurring can be derived in a similar way to those which have already been found for reflection at the boundary between two different media. The main differences from the previous case are that ρc is the same on both sides of the boundary, and that instead of the particle velocities being equal on both sides, the volume velocities are the same. Previously we considered a wave having a large wave front impinging on an even larger surface area. At the boundary itself there was no change in the area of the wave front normal to the surface, so that the condition of continuity merely required that the velocities normal to the boundary should be the same on both sides. Now we are considering a case where there is a change in the area of the wave front at the boundary itself, and we must modify the condition of continuity so as to keep the volume displacements equal.

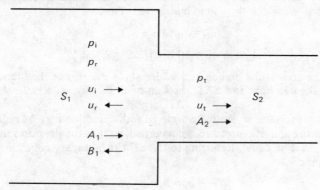

6.1 The notation for change of duct area

Using the notation in fig. 6.1 the boundary conditions are now

$$p_i + p_r = p_t$$

$$S_1 u_i + S_1 u_r = S_2 u_t$$

The second of these equations gives

$$(S_1/\rho c)p_i - (S_1/\rho c)p_r = (S_2/\rho c)p_t$$

Writing these out in full, and putting $x = 0$ and $t = 0$, leads to

$$A_1 + B_1 = A_2$$

$$A_1/S_2 - B_1/S_2 = A_2/S_1$$

If we compare this with the previous analysis, we see that instead of $\rho_1 c_1$ we now have S_2 and instead of $\rho_2 c_2$ we have S_1. There is no need to do the algebra all over again. We can write down the solutions straight away

$$\alpha_r = (S_1 - S_2)^2/(S_1 + S_2)^2$$

and

$$\alpha_t = 4S_1 S_2/(S_1 + S_2)^2$$
$$= (4S_1/S_2)/(1 + S_1/S_2)^2$$

For routine calculations it is convenient to express the sound power transmission coefficient in decibels, in exactly the same way as before

loss in dB = $10 \log (S_1/S_2)$ for high frequencies

and   loss in dB = $10 \log [(1 + S_1/S_2)^2/(4S_1/S_2)]$ for low frequencies.

6.3   Branch

At a branch we again consider two cases. At high frequency and short wavelength the total power divides in proportion to the branch areas with no reflection into the branch through which the sound has arrived. Then

$$\alpha_{th} = S_b/(\Sigma S - S_1)$$

loss in dB = $10 \log [(\Sigma S - S_1)/S_b]$

Where S_b is the area of the branch in question, S_1 is the area of the branch bringing the sound to the junction, and ΣS is the sum of all the branch areas including S_b and S_1.

At low frequency and so at long wavelength, the junction may be regarded as a point as far as the acoustic pressure is concerned, so that the pressures in all branches at a junction must be equal. Referring to fig. 6.2 the boundary conditions at a junction become

$$p_i + p_r = p_t = p_b$$

$$S_1 u_i + S_1 u_r = S_t u_t + S_b u_b$$

Where we use the suffices t and b for the transmitted and branch waves.

Using the complex notation and putting $t = 0$, $x = 0$ in the term $\exp [j2\pi(ft \pm x/\lambda)]$, we have

$$A_1 + B_1 = A_2 = A_3 \tag{6.1}$$

$$S_1 A_1/\rho c - S_1 B_1/\rho c = S_2 A_2/\rho c + S_3 A_3/\rho c$$

or

$$S_1 A_1 - S_1 B_1 = S_2 A_2 + S_3 A_3 \tag{6.2}$$

This gives three equations for the three unknowns B_1, A_2 and A_3. The same three

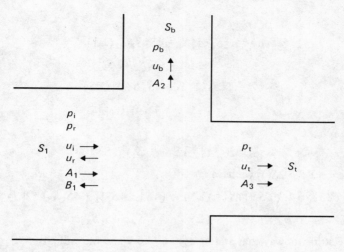

6.2 The notation for a duct with branches

equations would of course be obtained from the trigonometric notation once the relative phase angles had been found. The advantage of the complex notation lies in the fact that the phase as well as the magnitude is included in the complex coefficients.

Multiplying eqn 6.1 by S_1 and adding eqn 6.2 to eliminate B_1 gives

$$2S_1A_1 = S_1A_2 + S_2A_2 + S_3A_2$$
$$\therefore \quad A_2/A_1 = 2S_1/\Sigma S \tag{6.3}$$

The wave in the branch is also given by

$$A_3/A_1 = 2S_1/\Sigma S \tag{6.4}$$

To find the reflection coefficient, multiply eqn 6.1 by $(S_2 + S_3)$ and subtract eqn 6.2. Then

$$(S_2 + S_3) A_1 - S_1A_1 + (S_2 + S_3) B_1 + S_1B_1 = 0$$
$$\therefore \quad A_1 (S_2 + S_3 + S_1 - 2S_1) + B_1 (S_2 + S_3 + S_1) = 0$$
$$\therefore \quad B_1/A_1 = -(\Sigma S - 2S_1)/\Sigma S \tag{6.5}$$

To find the corresponding sound power transmission and reflection coefficients we use the relationship

$$I = p^2/\rho c$$

and

$$W = IS = p^2 S/\rho c$$

Here ρc is the same in all the branches, so it will cancel out, but we must take account of the different areas of the branches.

Then

$$\alpha_t = A_2^2 S_2 / A_1^2 S_1 = 4 S_1^2 S_2 / S_1 (\Sigma S)^2$$

$$= 4 S_1 S_2 / (\Sigma S)^2 \tag{6.6}$$

$$\alpha_b = A_3^2 S_3 / A_1^2 S_1 = 4 S_1^2 S_3 / S_1 (\Sigma S)^2$$

$$= 4 S_1 S_3 / (\Sigma S)^2 \tag{6.7}$$

and

$$\alpha_r = B_1^2 / A_1^2 = (\Sigma S - 2 S_1)^2 / (\Sigma S)^2 \tag{6.8}$$

If we expand $(\Sigma S - 2 S_1)^2$ we have

$$(\Sigma S)^2 - 4 S_1 \Sigma S + 4 S_1^2 = (\Sigma S)^2 - (4 S_1^2 + 4 S_1 S_2 + 4 S_1 S_3) + 4 S_1^2$$

$$= (\Sigma S)^2 - 4 S_1 S_2 - 4 S_1 S_3$$

We can see straight away that adding eqns 6.6, 6.7 and 6.8 will give

$$\alpha_t + \alpha_b + \alpha_r = (\Sigma S)^2 / (\Sigma S)^2 = 1$$

which is what we would expect, since it is implied in our assumption of no loss of energy.

In decibels, we have that the sound power transmission at a branch at lower frequency is

$$\text{loss in dB} = 10 \log [(\Sigma S)^2 / S_1 S_b]$$

This is another case where the intensity transmission coefficients are different from the power transmission coefficients, although the reflection coefficients are the same. The reason is once again that the area of the wave front has changed. Power ratios and intensity ratios are equal only if there is no change in the area of the wave front. If the areas do change, then the two ratios are related by

$$(W_2 / W_1) = (I_2 / I_1)(S_2 / S_1)$$

6.4   Expansion chamber

A very common low frequency silencing device, which is used both in high velocity air conditioning systems and in the simplest exhaust silencers, as well as for jet engine test cells, consists simply of an enlargement in a duct, fig. 6.3. Repeated reflections occur at each end of the expansion chamber, and an overall transfer coefficient can be found by the same kind of calculation that was used for transmission through a wall. In fact, there is no need to go through the whole of the mathematics all over again, for we have already seen that the difference between reflections in a duct and reflections between different media is that in a duct ρc remains constant but the change in area must be allowed for. In the most general case the equations contain the factors $\rho c / S$. With a change in medium but constant area the S cancels out leaving ρc; with a change in area in the same medium the ρc

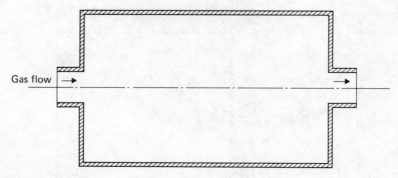

6.3 An expansion chamber

cancels out leaving $1/S$. Therefore we can write down the transfer coefficient by writing down eqn 5.23 with $1/S$ in place of ρc. This gives, for the expansion chamber, for which $S_3 = S_1$

$$\alpha_t = \frac{4/S_1^2}{(2/S_1)^2 \cos^2 lk_2 + (1/S_2 + S_2/S_1^2)^2 \sin^2 lk_2}$$

$$= \frac{4}{4 \cos^2 lk_2 + (S_1/S_2 + S_2/S_1)^2 \sin^2 lk_2}$$

This applies at low frequency. At high frequency the attenuation is negligible.

6.5   Helmholtz resonator

So far we have assumed that in all the reflected and transmitted waves the particle velocity is in phase with the acoustic pressure. This relationship was derived for a plane wave travelling without interruption through an endless medium, and we may use it so long as we ignore the effect at the junction of reflected waves originating elsewhere along the duct. It does not necessarily apply under other conditions.

Consider for example a closed volume V with rigid walls, and with a small opening of area S and length l connecting it to a much larger space, or to a duct. Such a cavity is known as a Helmholtz resonator, and we will show that at a certain frequency, depending on the volume of the cavity and the size of the opening, the sound pressure within the cavity is much larger than the incident sound pressure on the outside of the opening. This phenomenon is known as resonance and is analogous to the very large vibrations of a body mounted on a spring when it is excited by a periodic force at its resonance frequency.

We will denote the acoustic pressure outside the resonator by p_i and the acoustic pressure inside it by p_b, see fig. 6.4. For a perfect gas under adiabatic conditions,

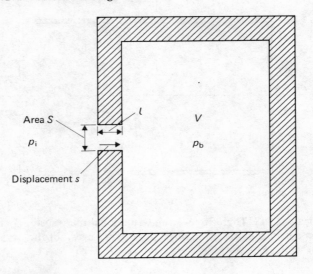

6.4 A Helmholtz resonator

the absolute pressure is related to the volume of gas inside the cavity by the equations

$$PV = wRT$$

$$\text{or} \quad P = \rho RT \tag{6.9}$$

$$\text{and} \quad PV^\gamma = \text{constant} \tag{6.10}$$

The small volume of gas in the neck of the opening moves bodily in and out of the cavity, so that the effective volume of the resonator cavity varies. The pressure variation accompanying this volume variation is the acoustic pressure p_b. If the displacement in the neck is s we can use eqns 6.9 and 6.10 to find a relationship between p_b and s. Thus from eqn 6.10

$$dP/P = -\gamma \, dV/V$$

$$\therefore \quad p_b = dP = -\gamma P \, dV/V$$

Substituting eqn 6.9

$$p_b = -\gamma \rho \, RT \, dV/V \tag{6.11}$$

But the velocity of sound for a perfect gas is $c = (\gamma RT)^{\frac{1}{2}}$

Substituting this in eqn 6.11 gives

$$p_b = -\rho c^2 \, dV/V \tag{6.12}$$

But the change in volume is due to the movements of the gas in the neck, or $dV = -sS$ the minus sign appearing because a positive value of s produces a reduction in V.

Using this value in eqn 6.12 we have

$$p_b = +\rho c^2 sS/V \qquad (6.13)$$

as the required relationship between p_b and s.

Now the force acting on the mass of gas in the neck is due to the difference between the pressure inside and outside the cavity acting on the area of the neck. We are considering the case in which p_i varies sinusoidally, so that $p_i = p_{i,0} \sin 2\pi ft$. Then

$$\text{force} = S(p_i - p_b) = Sp_{i,0} \sin 2\pi ft - \rho c^2 sS^2/V \qquad (6.14)$$

Since the force is equal to the mass times the acceleration, we may write

$$Sp_{i,0} \sin 2\pi ft - \frac{\rho c^2 sS^2}{V} = \rho lS \frac{d^2 s}{dt^2}$$

or re-arranging and dividing by S

$$\rho l \frac{d^2 s}{dt^2} + \frac{\rho c^2 sS}{V} = p_{i,0} \sin 2\pi ft \qquad (6.15)$$

This is the equation of motion for the gas in the neck of a Helmholtz resonator, assuming that this gas moves as a unit of constant mass, and ignoring all frictional and other losses.

The solution will be of the form

$$s = s_0 \sin 2\pi(ft + \phi) \qquad (6.16)$$

where s_0 and ϕ are the two constants of integration, whose value we must find by substituting known boundary values. Differentiating eqn 6.16 twice leads to

$$\frac{ds}{dt} = 2\pi fs_0 \cos 2\pi(ft + \phi)$$

and

$$\frac{d^2 s}{dt^2} = -4\pi^2 f^2 s_0 \sin 2\pi(ft + \phi) \qquad (6.17)$$

Substituting eqns 6.16 and 6.17 in eqn 6.15

$$-4\pi^2 f^2 \rho ls_0 \sin 2\pi(ft + \phi) + (\rho c^2 S/V) s_0 \sin 2\pi(ft + \phi) = p_{i,0} \sin 2\pi ft$$

$$\therefore \quad (\rho c^2 S/V - 4\pi^2 f^2 \rho l) s_0 \sin 2\pi(ft + \phi) = p_{i,0} \sin 2\pi ft$$

At $t = 0$,

$$(\rho c^2 S/V - 4\pi^2 f^2 \rho l) s_0 \sin 2\pi \phi = 0$$

$$\therefore \quad \phi = 0$$

so at $t = 1/4f$, $\sin 2\pi ft = 1$, and $(\rho c^2 S/V - 4\pi^2 f^2 \rho l) s_0 = p_{i,0}$

$$\therefore \quad s_0 = \frac{p_{i,0}}{\rho(c^2 S/V - 4\pi^2 f^2 l)}$$

Having thus found the two constants of integration, and incidentally confirmed that eqn 6.16 is a solution of eqn 6.15, we can substitute in eqn 6.16 to get

$$s = \frac{p_{i,0}}{\rho(c^2 S/V - 4\pi^2 f^2 l)} \sin 2\pi ft \qquad (6.18)$$

We must take the analysis a step further to find the relationship between p_b and p_i. Using eqn 6.13,

$$p_b = s\,\rho c^2 S/V$$

$$= \frac{c^2 S}{V(c^2 S/V - 4\pi^2 f^2 l)} p_{i,0} \sin 2\pi ft \qquad (6.19)$$

We see that when $c^2 S/V = 4\pi^2 f^2 l$ both the acoustic pressure in the resonator cavity and the movement of gas in the neck become infinite. Of course this is impossible, and what in fact happens is that as the amplitudes get larger the frictional and viscous resistances to the motion of the gas in the neck, which we have ignored, cease to be negligible and limit the actual amplitudes to large but nevertheless finite values. The effect on the value of the resonant frequency is small and we can say that the resonant frequency of a Helmholtz resonator is given by

$$2\pi f = \left(\frac{c^2 S}{lV}\right)^{\frac{1}{2}} \qquad (6.20)$$

In order to see the effect of using such a resonator as a side branch in a duct we need to find the relationship between the particle velocity at the neck and the incident pressure. We have

$$u = ds/dt = 2\pi f s_0 \cos 2\pi ft$$

$$= \frac{2\pi f}{\rho(c^2 S/V - 4\pi^2 f^2 l)} p_{i,0} \cos 2\pi ft$$

$$= \frac{1}{\rho(c^2 S/2\pi fV - 2\pi fl)} p_{i,0} \cos 2\pi ft \qquad (6.21)$$

The particle velocity is no longer in phase with the pressure, nor is the amplitude of their ratio constant, for it now varies with frequency. The phase difference can be more conveniently handled by means of the complex notation, which we can introduce by writing

$$p_i = p_{i,0} \exp j2\pi ft$$

Then eqn 6.18 becomes

$$S = \frac{p_{i,0}}{\rho(c^2 S/V - 4\pi^2 f^2 l)} \exp j2\pi ft$$

and the differentiation leading to the equivalent of eqn 6.21 is

$$u = \frac{ds}{dt} = \frac{j2\pi f}{\rho(c^2 S/V - 4\pi^2 f^2 l)} p_{i,0} \exp j2\pi ft$$

$$= \frac{\mathrm{j}p_{\mathrm{i}}}{\rho\,(c^2 S/2\pi f V - 2\pi fl)} \tag{6.22}$$

The ratio of acoustic pressure to volume displacement outside the resonator is then

$$p_{\mathrm{i}}/uS = \rho(c^2 S/2\pi f V - 2\pi fl)/\mathrm{j}S$$

$$= -\mathrm{j}\rho(c^2 S/2\pi f V - 2\pi fl)/S$$

$$= \mathrm{j}\rho(2\pi fl/S - c^2/2\pi f V) = \mathrm{j}X_{\mathrm{b}} \tag{6.23}$$

6.6   Resonator as a side branch

The previous section on the behaviour of Helmholtz resonators was introduced as a preliminary to seeing the effect of using such a resonator as a side branch in a duct. What we have done is to derive a relationship between pressure and volume displacement and we must use this relationship in a new equation to replace eqn 6.2. For the sake of simplicity we will put $p_{\mathrm{i}}/uS = \mathrm{j}X_{\mathrm{b}}$.

When resonant cavities are used in silencing devices, it is not usual to change the size of the main duct at the cavity, so we will take S_2 as being equal to S_1. Then the boundary conditions for which we will find the sound power transfer coefficient are

$$A_1 + B_1 = A_2 = A_3 \tag{6.24}$$

and

$$S_1 u_{\mathrm{i}} + S_1 u_{\mathrm{r}} = S_1 u_{\mathrm{t}} + S_3 u_{\mathrm{b}}$$

or

$$S_1 A_1/\rho c - S_1 B_1/\rho c = S_1 A_2/\rho c + A_3/\mathrm{j}X_{\mathrm{b}} \tag{6.25}$$

Multiplying eqn 6.24 by $S_1/\rho c$ and adding eqn 6.25

$$S_1 A_1 (2/\rho c) = S_1 A_2 (2/\rho c) + A_2/\mathrm{j}X_{\mathrm{b}}$$

$$\therefore \quad \frac{A_2}{A_1} = \frac{2S_1/\rho c}{2S_1/\rho c + 1/\mathrm{j}X_{\mathrm{b}}}$$

Multiplying top and bottom by $\mathrm{j}X_b \rho c/2S_1$

$$\frac{A_2}{A_1} = \frac{\mathrm{j}X_{\mathrm{b}}}{\mathrm{j}X_{\mathrm{b}} + \rho c/2S_1} \tag{6.26}$$

$$= \mathrm{j}X_{\mathrm{b}}\frac{(\rho c/2S_1 - \mathrm{j}X_{\mathrm{b}})}{(\rho c/2S_1)^2 + (X_{\mathrm{b}})^2}$$

$$= \frac{-X_{\mathrm{b}}^2}{(\rho c/2S_1)^2 + X_{\mathrm{b}}^2} + \frac{\mathrm{j}X_{\mathrm{b}}\rho c/2S_1}{(\rho c/2S_1)^2 + (X_{\mathrm{b}})^2}$$

$$= \alpha + \mathrm{j}\beta$$

The amplitude of $\frac{A_2}{A_1}$ is

$$\left|\frac{A_2}{A_1}\right| = (\alpha^2 + \beta^2)^{\frac{1}{2}} = \frac{[X_b^4 + (\rho c/2S_1)^2 X_b^2]^{\frac{1}{2}}}{(\rho c/2S_1)^2 + X_b}$$

$$= \frac{X_b[X_b^2 + (\rho c/2S_1)^2]^{\frac{1}{2}}}{(\rho c/2S_1)^2 + X_b^2} = \frac{X_b}{[(\rho c/2S_1)^2 + X_b^2]^{\frac{1}{2}}}$$

and the sound power transfer coefficient is

$$\alpha_t = \left|\frac{A_2}{A_1}\right|^2$$

$$= \frac{X_b^2}{(\rho c/2S_1)^2 + X_b^2} \qquad (6.27)$$

Since X_b is a function of the frequency and of the dimensions of the cavity, so also is the coefficient α_t. The value of the sound power transfer coefficient must be calculated for each frequency band for each particular resonator. But it can be seen that at the resonant frequency $X_b = 0$ and $\alpha_t = 0$. At this frequency the presence of the resonator causes all the sound energy arriving to be reflected back the way it came, and none is transmitted along the duct beyond the opening of the cavity.

Once again in order to produce usable results we have made a number of assumptions. The biggest one is that the gas in the neck moves as a solid lump or piston with relatively rigid ends. What actually happens at the ends of the neck is rather more complicated than this very simple picture suggests, but it can be allowed for by using an effective length for the neck, which is greater than the actual length. If a is the diameter of the opening, then the experimental results are that, depending on the exact shape of the restriction, the amount to be added to the actual length to arrive at the effective length lies between $0.6a$ and $0.85a$ at each end of the neck, making $1.2a$ to $1.7a$ as the total correction in most cases. However, $0.6a$ is the appropriate value in some cases.

One consequence of this empirical correction is that just a hole in the wall of a thin duct will act as the neck of a resonant cavity. Suppose such an opening communicates with a very large volume, such as the space between the duct and a larger enclosing duct, then in eqn 6.23 the term $c^2/2\pi fV$ will be very small compared with the term $2\pi fl/S$ and as a first approximation we can write

$$X_b = j\rho 2\pi fl/S$$

If the hole communicates with the atmosphere, some sound energy will be lost from the duct system by radiation to the outside, and another of our assumptions is no longer accurate. This loss of energy from the duct system can be allowed for by introducing a resistance term into eqn 6.15

$$\rho l \frac{d^2 s}{dt^2} + R_b \frac{ds}{dt} + \rho \frac{c^2 s S}{V} = p_{i,0} \sin 2\pi ft$$

It will be found that eqn 6.27 then becomes

$$\alpha_t = \frac{R_b^2 + X_b^2}{R_b^2 + (X_b + \rho c/2S_1)^2}$$

6.7   Reactive silencers

Silencers on motor vehicles and on guns consist of a combination of expansion chambers, branch ducts, and openings between concentric ducts. They all depend on the geometry of the duct setting up reflected waves, and in the frequency range for which the silencer is designed much of the energy is reflected back to the source. The principle on which they work is exactly the same as for the very simple cases which we have already discussed in detail. The difference is that there are many branches or openings, not necessarily all of the same size, close to each other, and close to the effective changes of section in the main duct. The reflections occurring at each one interfere with, or reinforce, the reflections occurring at each of the others, and a full theoretical analysis becomes extremely tedious, although no new principles are involved. Moreover, the number of possible arrangements is almost infinite, so that it is not practical to attempt a general analysis which would cover even a small number of the silencers found in practice. More specialized books must be consulted for the design data of actual silencers. Very effective silencers can be designed for the lower octave bands. A simple reactive silencer is shown in fig. 6.5.

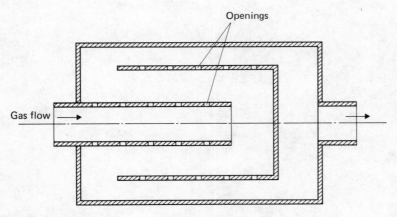

6.5 A reactive silencer

6.8   Bends

We have seen that at high frequency and therefore short wavelength, the duct dimensions are large compared with the wavelength, and the sound power divides more or less in proportion to the duct areas at branches and junctions. At low frequencies and therefore long wavelength, the duct dimensions are small compared with the wavelength, and a more detailed theory is necessary for what happens at branches and junctions. A small duct acts as a kind of wave guide and the sound power will follow changes of direction without much attenuation. In a large duct, however, the short wavelength energy will not be guided by the duct sides. When it reaches a bend it will impinge on the side of the bend and suffer attenuation by reflection.

The walls of ducts are not usually perfect, and will in general, like all walls, scatter the incident energy by reflecting it in all directions. With such diffuse reflection there will be a large number of reflections taking place from the wall of the bend, and it is not unreasonable to assume that equal amounts of energy are reflected along both directions of the duct, so that only half the incident energy continues along the duct after the bend. Halving the sound power transmitted past the bend means a reduction of 3 dB. Such a reduction is achieved at a given frequency by a bend in a duct whose dimensions are large compared with the wavelength, so at a given frequency it will occur in a large duct but not in a small one. In a given duct reduction will occur at a short wavelength or high frequency but not at a low frequency.

7 Transmission through the atmosphere

7.1 Free field transmission

At a point which is more than about two or three metres away from a source of sound, the numerical relations between acoustic pressure, particle velocity, energy density, and intensity, are the same for spherical waves as for plane waves. There remains, however, an important difference between spherical and plane waves which persists at all distances from the source; and that is the way in which both the pressure and the intensity vary with distance.

For a plane wave the area of the wave front remains constant, but for a spherical wave it increases proportionally with the square of the distance from the source. When this is allowed for in the theory it leads to the result that the pressure is inversely proportional to the distance from the source, and the intensity is inversely proportional to the square of the distance from the source. This is another example of the well-known inverse square law which appears in many other branches of physics. It should be noted that the inverse square law applies to the intensity, and of course the loudness, but not to the pressure, which is proportional to the square root of the intensity.

Where there are no surfaces to influence the way sound travels through the air it will propagate under free field conditions. A free field is simply any region of space in which sound is free to spread out in all directions. One important example is the **anechoic chamber**, fig. 7.1, of an acoustic laboratory where precautions are taken to make sure that the sound can spread freely in a relatively small area without suffering reflection. Such a chamber has highly absorbent walls so that no sound is reflected from them. The sound pressure can then be measured at a known distance from a source, and by integrating the intensity over a spherical area surrounding the source the total sound power radiated by the source can be found.

The other free field of importance is the open air, although in practice there are always trees, buildings, or other obstructions which will influence the overall transmission. Suppose for example that we are told that the sound level one metre from a small petrol engine is 70 dB(A).

Then the intensity is given by

$$10 \log (I/10^{-12}) = 70$$

$$\therefore \quad I/10^{-12} = 10^7 \, \mathrm{W \, m^{-2}}$$

$$\therefore \quad I = 10^{-5} \, \mathrm{W \, m^{-2}}$$

This is the intensity over the surface of a sphere of one metre radius. Therefore the power generated by the source is

$$4\pi \times 1^2 \times I = 4\pi \times 10^{-5} \, \mathrm{W}$$

7.1 An anechoic chamber (courtesy of the Building Research Establishment, Garston, Herts.)

or

$$SWL = 10 \log (4\pi \times 10^{-5}/10^{-12})$$

$$= 10 \log 4\pi \times 10^{7}$$

$$\therefore \quad SWL = 10 \times 8.1 = 81 \text{ dB}$$

Although this is a sound power rather than a sound pressure, the A weighting still applies to it, and must be used if a frequency analysis is required.

At a distance of say 10 metres, the intensity will have fallen to a fraction $1^2/10^2$ of its value at 1 metre, so the intensity will be

$$I_{10}/I_1 = 10 \log (1^2/10^2) = 10 \log 10^{-2}$$
$$= -20 \text{ dB}$$
$$\therefore \quad I_{10} = 70 - 20 = 50 \text{ dB(A)}$$

In general for a doubling of the distance we have

$$I_2/I_1 = 10 \log (1^2/2^2) = 10 \log (1/4)$$
$$= -10 \log 4 = -6 \text{ dB}$$

If, as is more likely because it is a more practical way of doing the test, the engine was standing on the ground and the pressure measurements taken over a hemisphere instead of a full sphere, then the acoustic power generated would be only $2\pi \times 10^{-5}$ watts, giving a SWL of 78 dB. The relationships between the intensities remain unaltered, and the SPL at 2 metres is still 64 dB(A) and at 10 metres 50 dB(A).

We can put these results into more general form. If the source generates W watts, its SWL is

$$\text{SWL} = 10 \log (W/10^{-12})$$

The intensity at a distance r metres is

$$I = W/4\pi r^2$$
$$\therefore \quad \text{SPL} = 10 \log (W/4\pi r^2 \times 10^{-12})$$
$$= 10 \log (W/10^{-12}) - 10 \log 4\pi r^2 \qquad (7.1)$$
$$= \text{SWL} - 10 \log 4\pi r^2$$
$$= \text{SWL} - 20 \log r - 11$$

If r is measured in feet, a correction factor must be used. Since one foot is 0.3 metres, the formula becomes

$$\text{SPL} = \text{SWL} - 10 \log 4\pi (0.3r)^2$$
$$= \text{SWL} - 10 \log 4\pi r^2 - 10 \log 0.09$$
$$= \text{SWL} - 10 \log 4\pi r^2 + 10 \log 11.11$$
$$= \text{SWL} - 10 \log 4\pi r^2 + 10.5$$
$$\therefore \quad \text{SPL} \approx \text{SWL} - 20 \log r$$

If the source is on the ground the sound energy is distributed over only half a sphere, and instead of eqn 7.1 we have

$$SPL = SWL - 10 \log 2\pi r^2$$

$$= SWL - 20 \log r - 8 \qquad (7.2)$$

In practical problems of transmission through the open air both the source and the receiver are usually sufficiently near the ground for eqn 7.2 to apply. The major exception is noise radiated by an aeroplane and heard on the ground. Sometimes one has to consider noise leaving a building by one window and re-entering through a different window. Here again the presence of the wall limits the free radiation to a hemisphere and eqn 7.2 applies.

Another way of looking at the difference between eqns 7.1 and 7.2 is that a large plane surface near the source reflects half the total radiated energy. A virtual image is formed on the other side of the plane, and when the radiation from the image is taken into account, the effect is to double the apparent strength of the actual source. The result is that the sound pressures at all points are 3 dB greater than they would be in the absence of the reflecting surface.

These expressions apply if the source radiates uniformly in all directions. If figures are quoted without any indication to the contrary we must assume that this is so. But many sources are known to have directional properties, i.e. they radiate relatively more power in one direction than in others. If a source of W watts, radiates uniformly in all directions then it will radiate $W/4\pi$ watts per unit solid angle in all directions. Suppose an actual source has a total power of W watts, but because it is non-uniform, in one particular direction it radiates w watts per unit solid angle. Then we define a directivity factor as

$$Q = \frac{w}{W/4\pi} = \frac{4\pi w}{W}$$

The intensity in this direction is now

$$I = w/r^2$$

and the SPL is

$$SPL = 10 \log (w/r^2)$$

$$= 10 \log (QW/4\pi r^2)$$

$$= SWL - 10 \log 4\pi r^2 + 10 \log Q$$

For a source on the ground the corresponding expression is

$$SPL = SWL - 10 \log 2\pi r^2 + 10 \log Q$$

The total power radiated in a cone of solid angle $d\theta$ is $w \, d\theta$ and the total power radiated is

$$W = \int_0^{4\pi} w \, d\theta$$

For a directional source, w is a function of θ and the integration cannot be carried out unless the nature of the dependence is known. For a uniform source w is constant and

$$W = 4\pi w$$

so that

$$Q = 1$$

7.2   Absorption

In a free field the sound energy spreads out so that the intensity falls off with the square of the distance, but the total acoustic energy over the whole wave front remains constant. Although the theory developed so far allows for changes in intensity along a particular path, we have all along assumed that there is no loss of total acoustic energy; it is merely divided up in various ways.

In the end, however, all acoustic energy is converted into heat. The loss of energy leads to a reduction in intensity, but the mechanism is totally different from any we have considered so far. This loss of energy applies to all sound waves, so that plane waves as well as spherical waves will suffer a loss of intensity due to the absorption of sound energy by the medium through which the sound is travelling. The absorption of sound energy by the transmitting medium is generally important only in the open air and sometimes in very large auditoria.

There appear to be three ways in which sound energy is converted into heat energy when sound is travelling through a fluid. The first is by means of viscous friction between the parts of the fluid which move relative to each other during the successive compressions and rarefactions. The heating effect due to this friction is obtained at the expense of the acoustic energy.

We have said that the successive compressions in a sound wave take place so rapidly that the process is adiabatic. There are therefore temperature changes in the fluid and differences in temperature occur between adjacent parts of it. Where there are temperature differences there must be a flow of heat. This is the second mechanism of sound dissipation. Energy which reaches part of the fluid in the form of acoustic pressure energy is transferred to an adjacent part in the form of thermal energy. As is to be expected, this effect is much more important in gases than in liquids.

In many fluids the attenuation due to viscous forces and thermal effects varies as the square of the frequency. In addition these two mechanisms affect the acoustic velocity, which also varies in many fluids as the square of the frequency. Not only is there a reduction in the amplitude of the vibration but there is also distortion of the sound pattern because components of different frequency travel at different velocities, so that at significant distances from the source the frequency spectrum is different from that at the source.

Because of the presence of both viscous and thermal damping the density does not change instantaneously with the pressure, but lags behind the pressure changes. The time lag is due to the finite time needed for energy to be transferred from one form to another. In some fluids the passage of a sound wave also involves the exchange of energy between the different forms of molecular energy of translation and vibration. These energy transfers also require a finite time to take place. They also lead to a time lag in the temperature changes which are associated with changes in internal energy, and are the third mechanism by which sound energy is converted to heat. The final effect of the time lags is once again that acoustic energy is permanently transformed into thermal energy and the amplitude of the sound wave is correspondingly attenuated.

Even where the absorption of sound energy is not entirely negligible it is very small, and it is a good approximation to assume that as the wave travels through a short distance dx the reduction in pressure due to absorption is a constant proportion of the pressure. This is the same as assuming that the energy absorbed is always the same fraction of the total energy. Then we may write

$$dp/p = -\alpha\, dx$$

$$\ln(p/p_0) = -\alpha x$$

$$\therefore \quad p = p_0\, e^{-\alpha x} \tag{7.3}$$

Since intensity is proportional to the square of the pressure, the corresponding equation for the intensity is

$$I = I_0\, e^{-2\alpha x}$$

We can express this as a decibel reduction,

$$\text{loss in dB} = 10 \log (I/I_0)$$

$$= 10 \log e^{-2\alpha x}$$

$$= 10 \times 0.434 \ln e^{-2\alpha x}$$

$$= -8.68\, \alpha x$$

The minus sign indicates that for positive values of α and x there is a loss and not a gain.

Attenuation is frequently measured in terms of nepers. A wave is said to be attenuated by one neper when its amplitude is reduced to $1/e$ of its initial amplitude. Then from eqn 7.3 a reduction of one neper occurs when $\alpha x = 1$, two nepers when $\alpha x = 2$, and so on. So the reduction in nepers is αx and in decibels 8.7 αx. The attenuation coefficient α is measured in nepers/metre.

7.3   Additional attenuation

The absorption in air is significantly affected by the presence of impurities and particularly of moisture, and it is found that the attenuation varies both with frequency and with humidity.

Particles of fog and smoke provide additional absorption in addition to that provided by the air itself, again by mechanisms which depend on the time taken for energy changes to take place. However, it appears that fog particles do not add a great deal of absorption above about 300 Hz. Moreover, whenever there is fog, there is generally less background noise due to traffic and other outdoor activities. When there is snow on the ground background levels also diminish. With a lower background level there is less masking, so that a given sound will be distinguished at lower intensities and so at greater distances from the source, although the actual attenuation is not altered. For calculation purposes it is perhaps best to ignore any attenuation due to rain, sleet, snow, or fog.

When sound travels through water the total absorption is also affected by any impurities or particles in the water. A similar effect takes place in water containing a large number of small bubbles, as well as providing additional absorption, the bubbles also scatter the incident sound wave. Although this attenuates the amplitude of the transmitted wave, it is not a true loss of acoustic energy, but may better be described as a wastage of energy; energy is lost from the main beam by being scattered in other directions but it remains in the form of acoustic energy, whereas absorption converts acoustic energy into thermal energy.

The presence of gas bubbles in a liquid, or of dust particles in a gas, alter both the bulk density and the bulk modulus of elasticity. Not only does this alter the velocity of sound through the fluid, but it leads to reflection and refraction wherever there is a change in the bulk properties of the fluid. Similar effects occur in a gas if there are temperature gradients in it, because the velocity of sound in a gas depends on the temperature, and we have seen that there is reflection and refraction wherever there is a change in the product ρc.

Both ρ and c vary with temperature and so in the earth's atmosphere ρc varies with height above ground level. The variation of ρc and its effect on sound transmission is not readily predictable, but its general effect is to deflect the sound waves downwards. This deflection can occur at quite great heights and at significant distances from the source. Such a deflected wave is known as a **sky wave**, fig. 7.2.

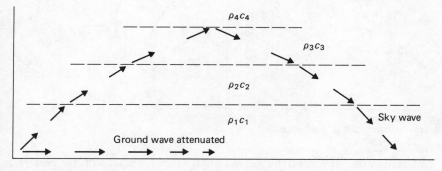

$\rho_4 c_4$

$\rho_3 c_3$

$\rho_2 c_2$

$\rho_1 c_1$

Sky wave

Ground wave attenuated

7.2 Refraction and reflection in the atmosphere

In certain atmospheric conditions sky waves make it possible for sounds to be heard at very great distances from the source while the ordinary processes of attenuation make the sound inaudible at intermediate points nearer to the source.

The variation of air temperature with height also promotes the existence of shadow zones. Normally the atmospheric temperature decreases with height, and since we have seen that the acoustic velocity is proportional to the square root of the temperature, the acoustic velocity also decreases with height. The effect is to distort the wavefront and deflect the sound upwards. Under these conditions a shadow zone may be formed right round the source, fig. 7.3.

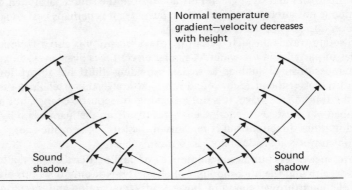

Normal temperature gradient—velocity decreases with height

Sound shadow

Sound shadow

7.3 Normal temperature gradient effect

In certain conditions, such as a clear night, the temperature gradient is inverted, and the temperature increases with height. In this case the acoustic velocity also increases with height and the sound wave is deflected downwards, fig. 7.4.

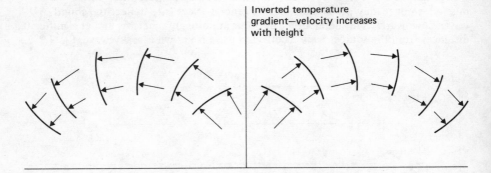

Inverted temperature gradient—velocity increases with height

7.4 Inverted temperature gradient effect

Unless both the source and the receiver are almost in contact with the ground, a sound wave can travel from the source to some intermediate point where it is reflected, so that at the receiver two waves arrive, one by direct transmission and one by reflection from the ground. Between 300 and 600 Hz the phase relations of the direct and reflected waves are such that the two waves sometimes interfere with each other and the total sound level is less than would have been expected. If the two waves are in phase with each other, then the total sound level is greater than that due to either wave alone.

While the ground may reflect some of the sound, it also acts as an energy absorbing surface. The absorption of energy by the ground adds further to the attenuation of sound waves travelling near the ground. Not surprisingly the amount of absorption depends on the nature of the ground, being slightly different for grass and shrub land than for forests.

Since sound is a vibration in the air the acoustic velocity, which depends on the properties of the air, is a velocity relative to the air. In the presence of wind the wind velocity must be added to the acoustic velocity to obtain a velocity relative to the ground. Even in a gale, the wind velocity is only about one tenth of the velocity of sound, so this effect alone would not make a great difference. Near the ground, however, the wind velocity almost always increases with height, and so the sound velocity, relative to the ground, increases with height downwind and decreases with height upwind. Some distance from the source the wavefront will have become distorted because of these differences in velocity. Now a wave is transmitted in such a way that the direction of propagation at any point is always at right angles to the wave front at that point. So the distortion of the wave front due to the variation in velocity leads to a deflection of the direction of propagation. In the downwind direction the sound is deflected downward. In the upwind direction the sound is deflected upward, and this upward deflection gives rise to an ill-defined **shadow zone** upwind from the source as illustrated in fig. 7.5.

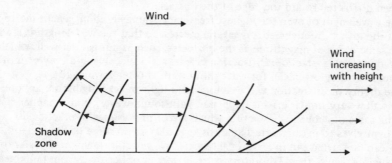

7.5 The effect of wind on wavefront

In summary, it can be said that the factors which influence the transmission of sound through the atmosphere are many and varied. For small distances the effects are not great and for calculation purposes can frequently be ignored. For longer paths they may need to be borne in mind but the longer the path the less certain does any calculation become.

7.4   Barriers and acoustic diffraction

Finally we must consider the effect of a solid barrier in the direct path of the sound. Let us imagine two people standing on opposite sides of a brick wall, the wall being high enough to block the direct line of sight so that they cannot see each other. If it is dark and one of them has a very bright lantern, the other one will not see the lantern, for the wall will cast a shadow, and the only light which reaches the eyes of the man standing in the shadow is such light as may be reflected from other objects in the vicinity. Although they cannot see each other they can hear each other fairly well. If one of them were completely enclosed in a brick building, they would not be able to hear each other. It appears that a free standing wall or barrier does not attenuate sound as much as if it forms part of a complete enclosure. The sound, unlike the light, can find its way round the top of the barrier or wall. Now, except where it is refracted by changes in the acoustic properties of the air, sound travels in straight lines, just as does light. How then does the sound manage to get round the wall when the light does not?

In fact it is not strictly accurate to say that waves travel in straight lines. Whenever a wave encounters an obstacle it is diffracted at the edge of the obstacle. One can rather loosely think of the wave as being bent round the obstacle. The amount of the bending, or the distance into the shadow zone for which the diffraction produces noticeable effects, depends on the dimensions of the object relative to the wavelength of the particular wave motion we may be considering. Since the wavelength of light is of the order of 10^{-7} metres, it is extremely small compared with all everyday objects, and this is why light appears to cast sharply defined shadows. But if the shadow is examined with a very powerful magnifying glass, it will be found that the edge of the shadow is not a sharp line and that the light has been diffracted round the edge of the obstacle.

The wavelength of even the highest frequency sounds is about a centimetre, and of the lowest frequencies it is several metres, so that the wavelength is always of the same order of magnitude as the obstacles, such as our garden wall, which are encountered in practice. Diffraction effects are noticeable and they cannot be ignored as they often can for wave motions of shorter wavelength.

The theory of diffraction was developed for light waves long before anyone realised that very similar effects happened with sound waves. It also happens that the cases of acoustic diffraction which are of importance in practice do not lend themselves to mathematical analysis as readily as do some of the experiments which can be carried out in an optical laboratory, and it is in any case easier to get reproducible results from laboratory experiments in optics than from field measurements in acoustics. For these reasons more is known about the diffraction of light than of sound. Nevertheless some of the results of optical diffraction have been applied to acoustics, and the theory has been tested with actual measurements.

In brief, a wall between a source and a receiver in the open, see fig. 7.6, does not

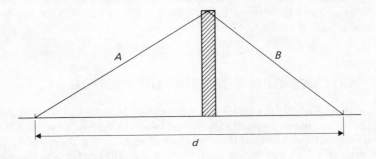

7.6 The notation for the diffraction of sound

block the sound in the way that one might at first expect from a solid barrier, but it attenuates the intensity at the receiver by an amount that depends on both the wavelength of the sound and on the difference between the distance which the sound has to travel in going over the top of the barrier and the straight line distance between source and receiver. If the distance from the source to the top of the barrier is A, from the receiver to the top of the barrier is B, and the shortest distance between source and receiver is d, then the attenuation depends on $(A + B - d)$ and on the wavelength λ.

When the shortest path of the sound over the top of the wall makes a fairly sharp angle at the top of the wall, the diffraction effect becomes less important than the flanking transmission paths which are always present in some form or other. For values of $2(A + B - d)/\lambda$ greater than about 10, reflection from the upper atmosphere and from other objects in the vicinity limits the total attenuation to between 20 and 24 dB. For values of $2(A + B - d)/\lambda$ less than about 1 the attenuation tends to a minimum value of about 5 dB.

In between these limits the attenuation varies approximately as

$$10 \log \left[2(A + B - d)/\lambda \right] + 13$$

or attenuation $= 10 \log (A + B - d) - 10 \log \lambda + 16$ dB $\qquad$ (7.4)

This semi-empirical formula may be used with confidence for values of $2(A + B - d)/\lambda$ between 1 and 10, for which it will give attenuations between 13 and 23 dB.

For a given height of barrier the attenuation will decrease as the distance between source and receiver increases. This is more or less what one would expect, for geometrically this is the same as reducing the height while keeping the distance the same.

Buildings form thick barriers for which diffraction occurs at both edges, that is at the roof line on both the source and receiver sides of the building. Consequently the attenuation may be higher than that given by the above formula. It has also been found experimentally that where there are a number of structures between a source and a receiver, it is the first structure which introduces the reduction in sound level; the presence of further structures does not give any further reduction.

In some cases the barrier can prevent the destructive interference of the direct and the ground reflected waves to which we have referred earlier. In these cases the attenuation at low frequencies is less than would be predicted, and in an extreme case the intensity at the receiver could even be higher with a barrier than without one.

The same considerations apply to a baffle placed at the end of a duct, for example where a ventilation duct terminates in a room. Eqn 7.4 applied between the end of a duct and a point in an average room shows that a very large baffle would be needed to achieve a significant reduction in sound level. In fact the attenuation will be less than the formula suggests because in this case diffraction occurs around four edges instead of only one. Moreover, in most rooms, as will be shown in more detail in Chapter 9, the sound pressure at any point is due more to repeated reflections from the walls than to the direct transmission from source to receiver. The baffle has no effect on the reflected energy, so a baffle at the end of a duct may be expected to have a negligible effect on the sound levels.

8   Absorption at walls and surfaces

8.1   Absorption coefficient

When we listen to a sound in a room, what we hear depends not only on the source of the sound, but almost as much on the area and nature of the walls and other surfaces present in the room. When the sound falls on a surface such as a wall some of the sound energy is reflected back, some is transmitted through the wall, and some is dissipated as heat in the material of the wall. The power of the sound is so small that even a large dissipation of sound energy has a negligible effect on temperature. To see that this is so let us consider a sound having the very high intensity of 100 dB and let us make the extreme assumption that the whole of the energy is converted into heat. From the definition of intensity we have

$$100 = 10 \log (I/10^{-12})$$
$$= 10 \log I + 10 \log 10^{12}$$
$$= 10 \log I + 120$$
$$\therefore \quad \log I = -2$$
$$\text{or} \qquad I = 10^{-2} \, \text{W m}^{-2}$$

We can equate the incident power to the rate of heat generation. If there were no heat loss from the wall the temperature would rise at a rate proportional to the rate of heat generation and inversely proportional to the mass and specific heat of the wall. Let us take a typical building partition having a density of about 2000 kg m^{-3}. If it is 0.1 m thick the mass per unit surface area is 200 kg m^{-2}. The specific heat will be about 0.9 kJ kg^{-1} °C^{-1} or 900 W s kg^{-1} °C^{-1} so temperature rise

$$= 10^{-2}/200 \times 0.9 \times 10^3$$
$$= 5.55 \times 10^{-8} \, °\text{C s}^{-1}$$

It is clear from this example that the order of magnitude of sound energy is so small in relation to other forms of energy that when sound is dissipated in the form of heat it can for all practical purposes be regarded as lost. For this reason the term sound absorption when used in problems concerned with room acoustics is given a wider meaning than when dealing with problems concerned with the transmission of sound through fluids.

We first define the **coefficient of sound reflection** as the ratio between the reflected sound energy and the incident sound energy. This is the same definition as has already been used when we were discussing reflection at discontinuities in ducts. Since it is impossible to reflect more energy than is received the sound reflection coefficient must always be less than one. We then define the **coefficient of sound absorption** α, as one minus the coefficient of sound reflection. This definition gives absorption a wider meaning than it has in everyday usage, for it includes transmission through the surface as well as real absorption at the surface. This is reasonable so long as we confine our attention to one side of the surface only. Thus when sound from a room reaches an open window practically all the energy passes out and a negligible amount is reflected back into the room. But, although there is no overall loss of sound energy, if we look at it from the point of view of anyone in the room, the opening must be treated as absorbing all the sound reaching it. It will be seen that the coefficient of absorption, as defined above, is in this case one.

Caution is needed in using measured values of absorption coefficients. The coefficient of absorption depends on a number of factors, and conditions in the field may be different from those under which it was measured in the laboratory. One of the relevant factors is how the material is used, for example how it is mounted. Plasterboard fixed directly to a brick wall will have a different value of α from plasterboard mounted on battens and then fixed to the wall.

The coefficient also varies with the frequency of the sound. Ideally separate values should be quoted for each frequency, but in practice simpler methods are used. Sometimes the value of α at a frequency of 500 Hz is used. Alternatively an average is taken of the values for 250, 500, 1000 and 2000 Hz. Either of these values may be referred to as the **noise reduction coefficient**.

Another important factor is that α depends on the angle at which the sound waves strike the surface. The actual value measured is an average for all the angles of incidence occurring under the conditions of the test, and it is important to realise that the value quoted may depend on the precise methods used in one particular test, and may not apply exactly elsewhere. If the distribution of the in-

8.1 A reverberation chamber showing an aperture for mounting test specimens (courtesy of the Building Research Establishment, Garston, Herts.)

cident sound is completely random then the coefficient is known as the **reverberant absorption coefficient.** An approximation to such completely random distribution is achieved in a **reverberation chamber,** fig. 8.1.

The word reverberation is used to describe the persistance of sound in an enclosure due to repeated reflections at the boundaries of the enclosure. A reverberation chamber contains a large number of irregularly shaped reflecting surfaces so that the sound is reflected from one surface to another until it is uniformly diffused throughout the room. In fact it is extremely difficult to achieve complete diffusion, although a very good approximation can be achieved in a well constructed reverberation chamber. Most test chambers give a greater degree of diffusion than is found in many ordinary rooms and offices, so care should be taken when applying test results outside the laboratory.

Absorption coefficients are frequently measured in a reverberation chamber. Using a certain source of sound, the sound level in the room is measured. A sample of the material is then introduced into the room, and the sound level with the same source of sound is again measured. The reduction in the measured sound level is due to the extra absorption of the sample of test material.

The sound absorption coefficient can also be calculated from the effect which a sample of the material has on the time rate of decay of sound energy in a reverberation chamber. This is the **Sabine absorption coefficient.** Measurements made in a reverberation chamber include the effects of diffraction occurring at the edge of the sample being tested, as well as any other boundary effects due to the finite size of the sample.

The other method of measuring an absorption coefficient is to mount a sample of the material at one end of a tube, with a small loudspeaker at the other end, and to measure the actual intensity of the sound reflected from the sample. This method, which is indicated in fig. 8.2 gives lower values than reverberation chamber methods for the same material.

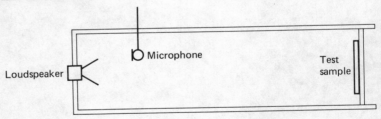

8.2 The measurement of the absorption coefficient

It can be seen that the conditions under which an absorption coefficient is measured are never the same as the conditions for which a calculation is required, and so some element of error is inevitable, and a degree of discretion is called for in using published data.

8.2   Equivalent absorption

In calculating the sound level in a room we have to take into account the absorption of all the surfaces. It is therefore convenient to use an average or mean absorption coefficient for the room. Because the amount of absorption is also proportional to area, the areas must also appear in the calculation of the mean absorption coefficient. The **mean absorption coefficient** is defined as

$$\bar{\alpha} = \frac{S_1\alpha_1 + S_2\alpha_2 + S_3\alpha_3 + \cdots}{S_1 + S_2 + S_3 + \ldots} = \frac{\Sigma S_i\alpha_i}{\Sigma S_i}$$

where S_1, S_2, S_3, etc., are the areas of surface having reverberant or Sabine absorption coefficients $\alpha_1, \alpha_2, \alpha_3$, etc., respectively.

Now the rate at which energy is absorbed by a surface is the product of the area, the absorption coefficient, and the rate at which energy reaches the surface. Another surface, having a different area and absorption coefficient, would absorb energy at the same rate relative to the incident energy providing the product of area and absorption coefficient were the same. This enables yet another method of expressing absorption to be used. Consider two surfaces which absorb energy at the same rate, if the absorption coefficient of one of them is unity, then its area is known as the **equivalent absorption** of the other surface. The equivalent absorption takes into account not only the coefficient of absorption but also the area of the absorbing surface. The definition can be extended to include all the walls, floor and ceiling of a room, even if they all have different absorption coefficients. The equivalent absorption of the room is simply the sum of the equivalent absorptions of each area. The definition can be further extended to cover not just the bare room, but all its contents.

The advantage of using the concept of equivalent absorption is that it can be measured and used in practice without a detailed knowledge of the coefficients of each of the many sound absorbing surfaces normally found in a room. It also enables us to make an allowance for the presence of people and furniture in a room without attempting the impossible task of calculating the surface area of such complicated objects. A number representing the equivalent absorption per person, multiplied by the number of people expected to be in the room at any time, is added to the equivalent absorption due to all other surfaces. A new average absorption coefficient for the room can be found by dividing the total equivalent absorption, including people and furniture, by the total surface area of the walls, floor and ceiling.

The formal definition of equivalent absorption, denoted by the letter a, states that it is that area of a surface having a reverberation absorption coefficient of unity, which would absorb sound energy at the same rate as the room or object. If the area is measured in square feet then the unit of equivalent absorption is known as the sabin. If the area is measured in square metres the unit is referred to as the metric sabin.

A little thought will show that equivalent absorption is in fact $a = S\bar{\alpha}$ where S is the total surface area ($S = S_1 + S_2 + S_3 + \ldots = \Sigma S_i$). For this reason it is sometimes referred to as the total absorption.

8.3 Absorbent materials

The acoustic properties of a room can be controlled by lining the walls and ceiling with material having a suitable absorption coefficient. Many materials are commercially available, the majority of them depending for their effectiveness on a sufficiently thick layer of porous material. The incident sound wave sets up a movement of air in the passages of the material. Since the passages are narrow the frictional resistance to flow is high and the kinetic energy of the flow in and out of the passages is converted by friction into heat. The degree of porosity, the thickness of the layer, and the frictional resistance to flow through the pores, will all affect the actual value of the acoustic absorption coefficient. The biggest difference between various materials is usually in the friction factor.

The requirement for an absorptive material is that it shall reflect as little as possible of the sound energy reaching it. This can be achieved with relatively large pores with low flow resistance. The bigger the air gaps relative to the solid material the more readily will the acoustic energy flow into the pores without reflection. But with large passages the friction is low, and the rate at which the energy having entered the pores is dissipated as heat is less. The acoustic flow energy which is not converted to heat before reaching the far surface of the material will be reflected there and will suffer further loss in its passage back to the front face. If it has still not been completely absorbed it will leave the material and will add to the energy initially reflected from the surface. So a high resistance to fluid flow means that little acoustic energy enters the absorptive layer, and a low resistance means that little of the entering energy is absorbed. Not only is the usual engineering compromise needed in the basic properties of the material, but a fairly thick layer of the material is essential. The thickness provides a sufficiently long passage so that the total frictional effect is large in spite of the low friction per unit length. If it is necessary to provide additional absorption in a room, it is rarely worth considering anything less than 2 cm thick, and thicknesses of 5 cm or more are not uncommon.

For a material with any given frictional flow resistance, there is a certain thickness which is enough to absorb almost the whole of the acoustic energy entering the pores. This particular thickness will give the maximum absorption, and a thicker layer of the same material will not provide any more absorption. At the optimum thickness all the entering energy is absorbed and the coefficient of absorption depends only on the initial reflection from the surface when the incident acoustic energy first strikes it.

The frictional loss depends on the velocity of flow. As the instantaneous particle velocity in an acoustic wave increases with increasing frequency, we might expect the coefficient of absorption to increase with frequency. This is indeed what is found in practice, although there is not usually a simple direct relationship. Highly compressed materials tend to have a high fluid friction and so a relatively low acoustic absorption.

It is necessary for the pores to form continuous channels, as in fibrous materials, for the air to flow through. There are synthetic foams in which the porosity is achieved by the formation of individual air pockets which are completely sealed from one another. Such a construction gives too high a flow resistance and is unsuitable for an acoustic absorbent.

Absorbent materials should not be painted, for the paint is liable to block the pores. Once the pores are blocked the whole of the absorbent properties are lost. The difficulty of decorating and cleaning a porous layer can be overcome by facing it with a perforated covering. Such a covering also provides useful mechanical protection for a soft fibrous material which could otherwise be easily damaged.

The protective surface is usually perforated to give about 13% free opening. It may be painted so long as the perforations are not clogged. Almost any material may be used for the covering panel, its effect is to increase the absorption at low frequencies but to reduce it at high frequencies. The increase at the low frequencies is extremely useful, because it is usually the low frequencies which are most difficult to deal with. The absorption at the high frequencies is usually large enough for the reduction not to matter.

If a material has a relatively high resistance to fluid flow its acoustic absorption can be improved by perforating it with a large number of cavities. These openings increase the effective surface area, and by presenting a larger total area to the flow of energy, reduce the effect of a high specific resistance. The shape of the openings is not important and a great variety are used in practice. A further advantage of such a construction is that the surface can be painted between the relatively large openings without fear of clogging the fairly small pores on which the acoustic performance depends. The earliest acoustic tiles of this form were made of sugar cane fibre, but one can now get other wood fibre panels, and most fibrous acoustic materials which are available as porous layers are also obtainable in the form of tiles.

8.4   Resonant absorption

When we considered a Helmholtz resonator in Chapter 6 we ignored all friction. In fact there is bound to be some frictional force resisting the motion of the gas in the neck of the opening. Let us make the usual assumption that friction is proportional to the velocity. If the constant of proportionality is β then eqn 6.15 becomes

$$\rho l \frac{d^2 s}{dt^2} + \beta \frac{ds}{dt} + \rho \frac{c^2 sS}{V} = p_{i,0} \sin 2\pi ft \tag{8.1}$$

If we again assume a steady state solution of the form

$$s = s_0 \sin 2\pi (ft + \phi)$$

so that

$$ds/dt = 2\pi f s_0 \cos 2\pi (ft + \phi) \tag{8.2}$$

and

$$d^2 s/dt^2 = -4\pi^2 f^2 s_0 \sin 2\pi (ft + \phi)$$

we will have

$$-\rho l\, 4\pi^2 f^2 \sin 2\pi (ft + \phi) + \beta 2\pi f \cos 2\pi (ft + \phi) + (\rho c^2 S/V) \sin 2\pi (ft + \phi)$$

$$= (p_{i,0}/s_0) \sin 2\pi ft$$

At $t = 0$,

$$(\rho c^2 S/V - 4\pi^2 f^2 \rho l) \sin 2\pi\phi + 2\pi\beta f \cos 2\pi\phi = 0$$

Dividing by $\cos 2\pi\phi$ gives

$$\tan 2\pi\phi = \frac{-2\pi\beta f}{\rho c^2 S/V - 4\pi^2 f^2 \rho l}$$

from which it follows that

$$\sin 2\pi\phi = \frac{-2\pi\beta f}{[(\rho c^2 S/V - 4\pi^2 f^2 \rho l)^2 + 4\pi^2 \beta^2 f^2]^{\frac{1}{2}}}$$

and

$$\cos 2\pi\phi = \frac{(\rho c^2 S/V - 4\pi^2 f^2 \rho l)}{[(\rho c^2 S/V - 4\pi^2 f^2 \rho l)^2 + 4\pi^2 \beta^2 f^2]^{\frac{1}{2}}}$$

At $t = 1/4f$,

$$(\rho c^2 S/V - 4\pi^2 f^2 \rho l) \sin(\pi/2 + 2\pi\phi) + 2\pi\beta f \cos(\pi/2 + 2\pi\phi) = p_{i,0}/s_0$$

$$\therefore \quad (\rho c^2 S/V - 4\pi^2 f^2 \rho l) \cos 2\pi\phi - 2\pi\beta f \sin 2\pi\phi = p_{i,0}/s_0$$

Substituting for $\sin 2\pi\phi$ and $\cos 2\pi\phi$ gives

$$p_{i,0}[(\rho c^2 S/V - 4\pi^2 f^2 \rho l)^2 + 4\pi^2 \beta^2 f^2]^{\frac{1}{2}}/s_0 = (\rho c^2 S/V - 4\pi^2 f^2 \rho l)^2 + (2\pi\beta fl)^2$$

$$\therefore \quad s_0 = p_{i,0}/[(\rho c^2 S/V - 4\pi^2 f^2 \rho l)^2 + 4\pi^2 \beta^2 f^2]^{\frac{1}{2}} \tag{8.3}$$

and

$$s = p_{i,0} \sin 2\pi(ft + \phi)/[\rho c^2 S/V - 4\pi^2 f^2 \rho l)^2 + 4\pi^2 \beta^2 f^2]^{\frac{1}{2}} \tag{8.4}$$

Now the energy dissipated in friction is equal to the frictional force times the velocity. We have taken the frictional force to be β times the velocity, so the energy loss is $\beta(ds/dt)^2$. From eqns 8.2 and 8.3 it will be seen that this is a maximum when $[(\rho c^2 S/V - 4\pi^2 f^2 \rho l)^2 + 4\pi^2 \beta^2 f^2]$ is a minimum. In most cases β is small, which was the justification for ignoring it before, and the maximum value of displacement and velocity occurs at a frequency very close to the natural frequency of the undamped Helmholtz resonator, $f = (c^2 S/4\pi^2 lV)^{\frac{1}{2}}$. In the region of this frequency the amplitude of both displacement and velocity becomes very high, and so therefore does the friction. A significant amount of the incident acoustic energy is then converted to heat, and the resonant cavity acts as an absorber.

Such absorbent cavities are mainly used at low frequencies, where it is difficult to get adequate absorption with porous layers. One advantage of using cavities to provide absorption is that they can be tuned fairly accurately to the frequency at which absorption is required, by choosing suitable dimensions for the cavities. A range of frequencies can be covered by a number of cavities of different sizes. Slight variations in the resonant frequency can be achieved by partly filling the cavity with sand or ashes, which also adds to the friction and so increases the absorption. The disadvantage of placing a number of absorbent openings in a wall is that it may be architecturally undesirable.

The frictional force on which the absorption depends can also be obtained from the bending resistance of a thin membrane. A flexible membrane with a small air gap behind it provides low frequency absorption in a manner very similar to that of a Helmholtz resonator. The elasticity of the air gap provides a force proportional to the displacement but in the opposite direction, in exactly the same way as the air inside the cavity; and the mass of the membrane acts in the same way as the mass of air in the neck of the resonator.

Let us consider a unit area of the membrane in fig. 8.3 placed a distance d in

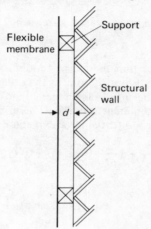

Flexible membrane

Support

Structural wall

d

8.3 A flexible membrane

front of a wall. The area corresponding to S in the previous equations is 1, and the volume V is $1 \times d$. The elasticity term which was $\rho c^2 s S/V$ therefore becomes $\rho c^2 s/d$. Instead of the mass term ρl we will have the mass of the membrane per unit area. Let us denote this by m. Then in place of eqn 8.1 we have

$$m \frac{d^2 s}{dt^2} + \beta \frac{ds}{dt} + \frac{\rho c^2 s}{d} = p_0 \sin 2\pi ft \qquad (8.5)$$

Instead of eqn 8.4 we have

$$s = p_0 / [\rho c^2/d - 4\pi^2 f^2 m)^2 + 4\pi^2 \beta^2 f^2]^{\frac{1}{2}}$$

The resonant frequency is given by

$$4\pi^2 f^2 m = \rho c^2/d$$
$$\therefore \quad f = (c/2\pi)(\rho/md)^{\frac{1}{2}} \qquad (8.6)$$

and this is very nearly the frequency at which the amplitude of vibration of the membrane is greatest. In the neighbourhood of this frequency the energy converted to heat in overcoming the bending resistance of the membrane is relatively great.

This analysis has ignored the elasticity of the membrane, which is usually very much less than that of the air gap. It has also assumed that the whole membrane moves bodily, or in phase, over an infinite area. In practice there are restraints on the movement of the membrane where it is fixed to the structure. Normally it will be glued or nailed to battens fixed to the wall, and the spacing of these will influence both the resonant frequency and the absorption achieved. Although this alters the numbers it does not alter the physical nature of the absorption process, and since it is not possible to calculate a numerical value of β the actual absorption coefficient must in any case be found experimentally.

Eqn 8.6 gives the resonant frequency with sufficient accuracy. For air at room temperature,

$$c\sqrt{\rho}/2\pi \approx 343 \times \sqrt{1.23}/2\pi \approx 60$$

and the maximum absorption is achieved at a frequency

$$f \approx 60(1/md)^{\frac{1}{2}}$$

where m is the mass of the membrane in $kg\,m^{-2}$ and d is the thickness of the air gap in metres. This frequency is usually low, of the order of 100 Hz. The usual value of the absorption coefficient of such a membrane is between 0.3 and 0.4.

The action of a flexible membrane can also be described in a slightly different way. If we use the complex exponential notation, eqn 8.5 becomes

$$m\frac{d^2s}{dt^2} + \beta\frac{ds}{dt} + \frac{\rho c^2 s}{d} = p_0\exp(j2\pi ft) \tag{8.7}$$

If

$$s = s_0\exp[j2\pi(ft + \phi)] \tag{8.8}$$

then

$$ds/dt = j2\pi fs$$

and

$$d^2s/dt^2 = -4\pi^2 f^2 s$$

Substituting in eqn 8.7 gives

$$(-4\pi^2 f^2 m + j2\pi\beta f + \rho c^2/d)\,s = p_0\exp(j2\pi ft) = p$$

Then at the membrane the particle velocity is

$$u = ds/dt = j2\pi fs$$

$$= \frac{j2\pi fp}{-4\pi^2 f^2 m + \rho c^2/d + j2\pi\beta f}$$

$$= \frac{jp}{\rho c^2/2\pi fd - 2\pi fm + j\beta}$$

$$\therefore\quad p/u = j2\pi fm - j\rho c^2/2\pi fd + \beta \tag{8.9}$$

As in the previous cases where we have used this notation, it provides a more succinct way of expressing the phase relationship between p and u than the trigonometric form.

When in Chapter 5, we considered the transmission of a plane wave past a boundary between two media, we used $\rho_2 c_2$ for the ratio p_t/u_t. If the boundary is formed by a flexible membrane, the value given by eqn 8.9 is more appropriate for this ratio. Now inspection of eqn 5.7 will show that when $\rho_2 c_2 = \rho_1 c_1$ there is no reflection, and therefore complete absorption in the wider sense of the word. It is theoretically possible that this could be achieved with a suitable membrane. The required condition is

$$j(2\pi fm - \rho c^2 / 2\pi fd) + \beta = \rho c$$

This in turn requires that

$$2\pi fm = \rho c^2 / 2\pi fd$$

or

$$f = (c/2\pi)(\rho/md)^{\frac{1}{2}}$$

that is, zero reflection can occur only at the resonant frequency. If at this frequency β is equal to ρc then there will be no reflection and the membrane will act as a perfect absorber. Under these highly idealised conditions the whole of the incident acoustic energy would be converted into heat by the internal friction of the membrane.

The theory could be extended to find the coefficients of reflection and absorption at any frequency, but it would not help much because of the difficulty of finding reliable values of β. It should be noted that we have assumed that the wavelength is large compared with the thickness of the gap, and that the backing wall is very much stiffer than the membrane. If the gap is large it is more appropriate to treat the membrane as a partition and apply the theory of Chapter 5.

8.5   Other absorbers

Perhaps the most useful absorptive treatment is a suitable porous layer about 5 cm thick, backed with a solid but flexible material, and mounted 3 to 5 cm in front of the wall. The porous material provides good absorption at high frequencies, and the presence of the air gap provides extra absorption at low frequencies. A variety of such materials, with various surface finishes, is available with absorption coefficients of 0.7 or more from 50 Hz to 4000 Hz. In some cases the surface finish is itself a thin flexible membrane which is almost transparent to high frequency sound and provides the additional absorption required at low frequencies.

Clothing and upholstery have fairly good sound absorbing properties, and the presence of people and seating in a theatre, or concert or assembly hall, has a significant effect on the total absorption. For practical purposes it is impossible to estimate the area of such surfaces, so the concept of total absorption must be used for calculation purposes. The absorption is widely variable, depending on the amount and nature of clothing worn and on its aspect in relation to the direction from which the sound is coming. The absorption per person can be anything between 2 and 5 sabins. The absorption of tip-up seats may depend on whether the seats are in the up or down position, although this dependence can be reduced by perforating the underside of the seats so as to increase the absorption when they are tipped up. The total absorption will vary with the number of people present.

However, a person sitting in a seat blocks at least part of the absorption of that seat, so that the total absorption of a deeply upholstered seat may be nearly the same whether it is occupied or unoccupied.

8.6   Absorptive silencers

Absorption occurs not only at the walls of rooms, but also at the walls of ducts. The most important example being that of ventilation ducts. The absorption of the duct walls reduces the intensity of the sound as it travels down the duct. The degree of attenuation depends on the amount of duct wall, and also on the cross sectional area of the duct, as well as on the coefficient of absorption. An empirical relationship has been found for the attenuation per unit length of duct.

$$\text{Attenuation} = 3.5\,\alpha^{1.4}P/S \text{ dB per metre} \tag{8.10}$$

where α is the coefficient of absorption, P is the perimeter of the duct in metres, and S is the cross sectional area of the duct in square metres.

The ratio P/S is the reciprocal of the hydraulic mean diameter, which appears in certain problems in fluid flow. For geometrically similar cross sections, the larger duct will have a lower value of P/S. For good attenuation it can be seen that P/S should be as large as possible. This is unfortunate, because for good airflow and low resistance it should be as small as possible. A large duct will give little acoustic attenuation but will have little hydraulic resistance. A small duct will have better acoustic attenuation but more hydraulic resistance.

The shape of the duct also affects the value of P/S. The minimum value for a given area is a circle, which is therefore the worst shape if silencing is required but the best shape for a speaking tube. A close runner-up is the square. It is possible to increase the ratio P/S of a square or rectangular duct without altering the overall dimensions by using splitters. This is the same as replacing one duct by one or more thinner and wider ducts which have a greater attenuation. However, this can only be done at the cost of increasing the hydraulic resistance.

If some ventilation ducting connects a sound source in one room to a different room, it very often happens that more attenuation is required than is provided by the ducting alone. The duct attenuation can be increased by lining the duct with an acoustic absorbent having a high value of α. It is not always necessary to line all the duct sides. If different sides have different values of α then an average coefficient of absorption must be used in eqn. 8.10, given by

$$\alpha = \Sigma P_n \alpha_n / \Sigma P_n \quad (n = 1, 2, 3, \ldots)$$

where P_n are the lengths of those sections of perimeter having coefficients of absorption α_n respectively.

It is usually better and more effective to insert an absorbent type silencer at a suitable part of the ducting. Silencers are commercially available to suit most applications, and there is little difficulty in getting special ones made for particularly difficult requirements. They consist of a length of ducting with splitters, the whole of the ducting and splitters being lined with thick sound absorbent material protected by a perforated covering. Suitable materials are available so that silencers can be used in many corrosive and high temperature atmospheres. A typical absorbent silencer is shown in fig. 8.4.

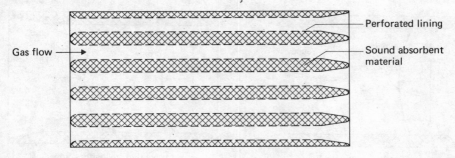

8.4 An absorptive silencer

The acoustic performance of a silencer can be expressed in two ways. The commonest and more useful is the **insertion loss** which is defined as the difference in decibels between the sound pressure levels or sound power levels at a point in the ducting before the silencer is inserted and after it is inserted. It is a difference between measurements made at the same point in space but at different times, a change having been made to the ductwork in the interval between the two measurements. The insertion loss can be added arithmetically to the attenuation of the rest of the ducting. Alternatively, the required insertion loss can be found as the difference between the overall attenuation which it is desired to achieve and the attenuation which is available without the silencer.

Noise reduction or SPL difference is the difference between the SPLs at the inlet and outlet sides of the silencer. It is a difference between measurements made at the same time but at different points in space. It is much more easily measured than insertion loss but is not so useful in design calculations, because it does not give a direct comparison with sound levels in the absence of a silencer.

We have seen that reflections occur at any change of section, shape or direction in a duct. Because of these reflections the presence of the silencer will affect the sound pressure on the acoustically upstream and the acoustically downstream side of the silencer. The noise reduction does not allow for the slight effect that the silencer has on the SPL on the inlet side, and it will usually be found that the measured value of noise reduction is a little greater, perhaps by about 3 dB, than the measured value of the insertion loss.

The performance of a silencer is affected by the flow of fluid through it. The velocity of sound is a velocity relative to the fluid stream. If the sound is travelling with the flow, its velocity relative to the silencer is greater than if the sound is travelling against the flow. In the latter case there is a longer time during which effective absorption can take place, and the attenuation will be correspondingly greater. How big the effect is depends on whether the boundary layer of the flow through the silencer is laminar or turbulent.

At any cross section the velocity of fluid flow is not uniform across the silencer, but is higher at the centre of the stream than at the boundaries. The result is a refraction or bending of the sound waves in the same way as that produced in the open air by variations of wind velocity. If the sound is travelling with the fluid it is refracted towards the boundaries, as shown in fig. 8.5. As more acoustic energy

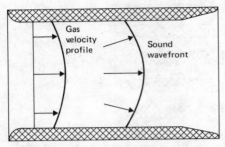

(a) Sound travelling with gas flow

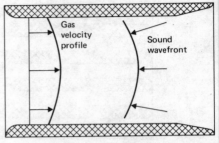

(b) Sound travelling against gas flow

8.5 Refraction in a silencer

is brought into contact with the absorbent material the attenuation will increase. If the sound is travelling against the fluid flow it will be refracted away from the boundaries and the attenuation will decrease. Quantitatively the effect of refraction is important only at high frequencies where the wavelength is smaller than the duct size. Because of the effect of fluid flow the only way to be certain of silencer performance is to test one under conditions resembling as closely as possible those in which it is expected to be used, and most silencer manufacturers now include experimental test data in their catalogues.

The insertion loss of absorptive silencers increases with frequency, as would be expected from the increase of α with frequency, and they should be regarded as complementing reactive silencers which are more effective at low frequencies.

9 Sound in an enclosure

9.1 Diffuse sound field

Sound waves in a room are reflected many times from surface to surface, with some absorption occurring at each reflection. Fig. 9.1 shows a few of the many sound paths between a source and a receiver. Because of these multiple reflections

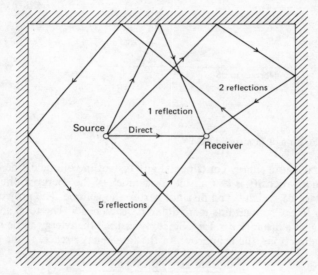

9.1 Sound paths in an enclosure

the sound pressure at any point in the room is greater than it would be at the same distance from the source in a free field. The contribution due to the reflections is known as the reverberant field and its importance can be shown by considering what happens when a sound dies away in a room. If a pistol is fired in an empty theatre, the sound reverberates throughout the auditorium, and can still be heard a significant time after the instant of the explosion.

Thus the sound in a room can be regarded as being made up of two parts. One is the sound which reaches a point directly from the source, just as it would in a free field. It obeys the free field laws of propagation, and in particular, the inverse square law. The other is the reverberant sound which, if there are a sufficiently large number of reflections before all the energy is absorbed, reaches any one point from all directions. In such a diffuse field, with sound travelling uniformly in all directions, the relationship between sound pressure level and the energy reaching a surface is no longer the same as for a plane wave, and we must consider the energy relations of diffuse sound in more detail.

We will treat the waves as plane waves for which the acoustic pressure and the particle velocity are in phase. The whole of the energy density then contributes to the travelling wave and we can find the intensity by the same kind of reasoning as was used for plane waves. Fig. 9.2 shows the geometry involved.

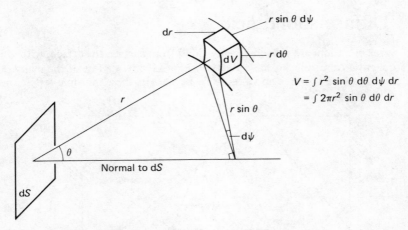

9.2 The notation for a diffuse sound field

Let the total sound energy contained at any instant in volume dV be $\mathcal{E}\,dV$ so that the average energy density is $\mathcal{E}$. Consider how much of the energy in the volume dV reaches a surface dS which is at a distance r from the volume. The projected area perpendicular to the straight line from dV to dS is $dS \cos\theta$. The total area surrounding the volume at a distance r is $4\pi r^2$. Since we assume the energy to be radiated equally in all directions, the proportion of the total energy reaching the surface is

$$\frac{\text{projected area of } dS}{\text{total area}} = \frac{dS \cos\theta}{4\pi r^2}$$

and the total energy reaching dS from dV is

$$\Delta E = \frac{\mathcal{E}\, dV\, dS \cos\theta}{4\pi r^2}$$

Now let us go on to consider how much energy reaches the surface dS from all infinitesimal volumes at the same distance r. This is done by integrating with respect to V. Since all elements of V are to be at the same distance from the infinitesimal surface, we are in fact considering a hollow spherical shell for which

$$dV = 2\pi r \sin\theta\, r\, d\theta\, dr$$

$$= 2\pi r^2 \sin\theta\, d\theta\, dr$$

Since we are mainly interested in the walls of the enclosure, and in the surfaces of solid objects in it, we can consider sound reaching the surface from one side only. The total volume is then found by integrating with respect to θ from 0 to $\pi/2$.

$$\Delta E = \int_0^{\pi/2} \frac{\mathscr{E} \, dS \cos \theta}{4\pi r^2} \, 2\pi r^2 \sin \theta \, d\theta \, dr$$

$$= \frac{\mathscr{E} \, dS \, dr}{2} \int_0^{\pi/2} \sin \theta \cos \theta \, d\theta$$

$$= \frac{\mathscr{E} \, dS \, dr}{4}$$

This is the amount of energy which will arrive at the surface in the time taken for sound to travel through the thickness of the hemispherical shell. Thus the time taken for this amount of energy to arrive at dS is

$$\Delta t = dr/c$$

Therefore the rate of energy arriving at dS is

$$\frac{\Delta E}{\Delta t} = \frac{\mathscr{E} \, dS \, dr}{4} \frac{c}{dr} = \frac{\mathscr{E}c}{4} \, dS$$

Now intensity is the rate of energy arrival per unit are, so

$$I = \mathscr{E}c/4 \qquad (9.1)$$

We see that the intensity of a perfectly diffuse sound at a surface is one quarter of the intensity due to a normally incident plane wave having the same energy density. In other words, to produce the same intensity, a diffuse wave arriving uniformly from all directions in a hemisphere on one side of a surface, must have four times the energy density of a plane wave arriving normally to the surface. In Chapter 2 we found a relationship between the intensity and the energy density for the case of a plane wave arriving at a surface at right angles to the surface. We have now found the corresponding relationship for the random arrival of sound from all directions on one side of the surfaces which set up the reverberant field.

The relationship between energy density and pressure is however the same as before. The direction of travel does not enter into this relationship, and our previous analysis applies equally well to sound travelling randomly in all directions. We therefore have

$$\mathscr{E} = p^2/\rho c^2$$

$$\therefore \quad I = \mathscr{E}c/4 = p^2/4\rho c$$

Putting ρc and 415 rayls and dividing by $I_0 = 10^{-12}$, gives

$$I/10^{-12} = (p/20.37)^2/4 \times 10^{-12}$$

$$= (p/2 \times 10^{-5})^2 (1/1.019)^2/4$$

$$10 \log (I/10^{-12}) = 20 \log (p/2 \times 10^{-5}) - 20 \log 1.019 - 10 \log 4$$

$$\therefore \quad \text{IL} = \text{SPL} - 0.0168 - 6$$

$$\approx \text{SPL} - 6 \qquad (9.2)$$

Eqn 9.2 is the practical form of eqn 9.1. The minus 6 dB in eqn 9.2 has the same significance as the 1/4 in eqn 9.1; to say that the intensity of a diffuse sound at a surface is 1/4 of the intensity of a normally incident plane wave is to say that the intensity level is 6 dB less. In Chapter 2 it was shown that the intensity level of a plane wave is approximately equal to the sound pressure level, and we have now shown that for a diffuse sound arriving from one side of a surface it is 6 dB less.

9.2   Sound decay

The rate at which sound decays in a room once the source has stopped radiating direct sound energy provides a useful numerical description of the acoustic properties of the room. Once the direct sound is zero the whole of the sound is due to the reverberant field, which will gradually die out due to absorption at the walls.

We find the decay rate by considering the rate of energy absorption. The rate at which energy arrives at a surface, multiplied by the absorption coefficient of the surface, is the rate at which energy is being removed at the surface. Now the rate at which energy arrives is the intensity multiplied by the area. So the rate of energy removal is $\alpha_1 I S_1$, or averaged over the whole room, $S\alpha I$.

The total energy in the room is the energy density multiplied by the volume, so the rate of energy removal must also be the product of the volume V and the time rate of change of energy density, a decrease being indicated by a negative sign for the rate of change.

Equating these two expressions for the rate of energy decrease gives

$$-V\frac{d\mathcal{E}}{dt} = S\alpha I \qquad (9.3)$$

But

$$\mathcal{E} = \frac{4I}{c}$$

so

$$\frac{d\mathcal{E}}{dt} = \frac{4}{c}\frac{dI}{dt}$$

and

$$-\frac{4V}{c}\frac{dI}{dt} = S\alpha I = aI$$

$$\therefore \quad \frac{dI}{dt} = \frac{-ac}{4V}I$$

$$\therefore \quad \frac{dI}{I} = \frac{-ac}{4V}dt$$

$$\therefore \quad \ln I = \frac{-ac}{4V}t + \text{constant}$$

If the initial intensity at time $t = 0$ is I_0 then the constant of integration is $\ln I_0$ and we can write

$$\ln I - \ln I_0 = -(ac/4V)t$$

$$\ln (I/I_0) = -(ac/4V)t \tag{9.4}$$

$$I/I_0 = \exp\left[-(ac/4V)t\right] \tag{9.5}$$

9.3 Reverberation time

We see that the rate at which sound decays in a room depends on both the total absorption and on the volume of the room. The rate of decay of sound intensity is an important property of a room. Differences in the rate of decay go a long way to explain why the same sound source gives a different impression in different rooms. The conventional way of expressing this property is by means of the reverberation time. **Reverberation time** is defined as the time taken for the intensity or the sound pressure level to fall to 60 dB of its initial value. A reduction of 60 dB means the sound energy has fallen to one millionth of its original value, or from a fairly loud sound to about the threshold of audibility.

The reverberation time T can be found by putting $10 \log (I/I_0)$ in eqn 9.4 equal to -60, then t becomes by definition equal to T. We have

$$-60 = 10 \log (I/I_0)$$

$$= (10/2.3) \ln (I/I_0)$$

$$= -(10/2.3)(ac/4V)T$$

$$\therefore \quad T = 55.2 \ V/ac$$

Putting $c = 343 \ \mathrm{ms^{-1}}$, gives, in SI units,

$$T = 0.161 \ V/a \ \mathrm{s} \tag{9.6}$$

In British units, $c = 1130$, ft s^{-1}

$$T = 0.049 \ V/a \ \mathrm{s}$$

This is the **Sabine formula** for reverberation time.

Slightly different formulae for the reverberation time can be found by making different assumptions about the nature of the energy absorption process. Instead of using a steady rate of energy removal proportional to the intensity, we can calculate the average time taken for the sound energy to travel across the room between reflections and the amount of energy removed at each successive reflection.

The total energy in the room at any instant is $\&V$. Since the intensity at the surface is $\&c/4$, the rate at which energy falls on the whole interior surface of the room S is $\&cS/4$. The time required for the whole of the energy to reach the surface is the total energy divided by the rate at which energy reaches the surface, or

$$\delta t = \frac{\&V}{\&cS/4} = \frac{4V}{cS} \tag{9.7}$$

This is the average time for one reflection to occur. The distance travelled by sound in this time is the mean free length

$$\delta l = c \, \delta t = 4V/S \tag{9.8}$$

The proportion of energy absorbed by one reflection is $\bar{\alpha}$, and the proportion left after reflection is $(1 - \bar{\alpha})$. After n reflections the total energy left is $\mathcal{E}_0(1 - \bar{\alpha})^n$. If one reflection takes place, on average, in $4V/cS$ seconds, then n reflections take $4Vn/cS$ seconds.

So we have

$$\mathcal{E} = \mathcal{E}_0(1 - \bar{\alpha})^n$$

and

$$t = (4V/cS)n$$

$$n = (cS/4V)t$$

$$\mathcal{E}/\mathcal{E}_0 = I/I_0 = (1 - \bar{\alpha})^{(cS/4V)t} \tag{9.9}$$

This can be turned into a more usual, and more useful form, by noting that

$$x^y = \exp(y \ln x)$$

So we may write

$$I/I_0 = \exp[(cSt/4V)\ln(1 - \bar{\alpha})]$$

To get the reverberation time we take logs to the base 10 and as before $10 \log(I/I_0) = -60$,

$$\therefore \quad -60 = 10 \log \exp[(cST/4V)\ln(1 - \bar{\alpha})]$$

Now

$$\log e = 0.434$$

so

$$10 \log e^x = 4.34x$$

so

$$-60 = 4.34(cST/4V)\ln(1 - \bar{\alpha})$$

$$T = -[60 \times 4V/4.34 \times cS \ln(1 - \bar{\alpha})]$$

and putting $c = 343 \text{ ms}^{-1}$,

$$T = \frac{0.161 \, V}{-S \ln(1 - \bar{\alpha})} \tag{9.10}$$

which is the **Norris–Eyring formula**.

In devising this expression we have used an average absorption coefficient $\bar{\alpha}$. It would perhaps be more accurate to use an average value for the proportion of energy left after one reflection, that is, an average value for $(1 - \bar{\alpha})$. This makes no difference so long as we use an arithmetic mean, but if we use a geometric mean value, then

$$(1 - \alpha)_m^S = (1 - \alpha_1)^{S_1} \times (1 - \alpha_2)^{S_2} \times \ldots.$$

$$= \Pi(1 - \alpha_i)^{S_i}$$

$$\text{and } (1 - \alpha)_m = \Pi(1 - \alpha_i)^{S_i/S}$$

We then have instead of eqn 9.9

$$I/I_0 = [\Pi(1 - \alpha_i)^{S_i/S}]^{(cS/4V)t} = [\Pi(1 - \alpha_i)^{S_i}]^{ct/4V}$$

$$= \exp\{(ct/4V)\ln[\Pi(1 - \alpha_i)^{S_i}]\}$$

$$= \exp\{(ct/4V)\ln[(1 - \alpha_1)^{S_1} \times (1 - \alpha_2)^{S_2} \times \ldots.]\}$$

$$= \exp\{(ct/4V)[S_1 \ln(1 - \alpha_1) + S_2 \ln(1 - \alpha_2) + \ldots.]\}$$

$$= \exp\{(ct/4V) \Sigma [S_i \ln(1 - \alpha_i)]\}$$

The reverberation time now becomes

$$T = \frac{0.161 \; V}{\Sigma [-S_i \ln(1 - \alpha_i)]} \tag{9.11}$$

This is the **Millington–Sette** formula. It is of value when there is a big variation in the values of the absorption coefficients of the different surfaces, for it includes the individual values for each surface. But the Sabine formula does the same if one writes is as

$$T = \frac{0.161 \; V}{\Sigma S_i \alpha_i} \tag{9.12}$$

The Sabine formula was derived by considering a continuous process, whereas the Millington–Sette formula was derived by considering a series of discrete reflections. The latter may be a more accurate picture of what actually happens, but the validity of the mathematics depends on the validity of using a statistical mean free path and the average time between reflections associated with it.

In any case the value found for the reverberation time depends on the value chosen for the absorption coefficients, and the answer can be no more accurate than the assumptions made in choosing suitable values. Theoretical rigour is less important than consistency in making the same assumptions for all calculations. Apart from its simplicity, the Sabine formula has the very real advantage that many published values of absorption coefficients were found by applying the formula to decay rates measured in test rooms containing a sample of the material being tested. It is therefore reasonable to suppose that the same values used in the same formula will give a reliable answer when applied to rooms other than the test room.

We can also show that for small values of the average absorption coefficient the Norris–Eyring formula approximates to the Sabine formula, by expanding $\ln(1 - \bar{\alpha})$ as an infinite series.

$$\ln(1 - \bar{\alpha}) = -\bar{\alpha} - \bar{\alpha}^2/2 - \bar{\alpha}^3/3 - \ldots. - \bar{\alpha}^n/n - \ldots.$$

This series is convergent for all values of $\bar{\alpha}$ between -1 and $+1$ and so for all practical values of $\bar{\alpha}$. If $\bar{\alpha}$ is less than 0.1, the absolute error in ignoring all terms except the first is less than about 0.005, or less than about 5% of the correct value.

The same expansion applied to the Millington—Sette formula shows that it, too, will give the same value of reverberation time unless at least one of the absorption coefficients is much greater than 0.1, because for small values of α_i the sum of $S_i \ln (1 - \alpha_i)$ is approximately equal to the sum of $S_i \alpha_i$.

The formulae all give the same answer for small coefficients of absorption. At very high values of α the Sabine formula cannot be used, because it gives a finite value of T for $\alpha = 1$. When $\alpha = 1$ there is no reverberant field and the reverberation time T is indeterminate. Both the other formulae are consistent with the physical requirement.

The main conclusion from this discussion must be that the calculation of a theoretical reverberation time is never exact.

9.4   Rate of sound increase

If we consider a sound increasing from zero intensity to a final steady intensity I_1, the rate of energy increase plus the rate of energy absorption must be equal to the steady rate of energy supply from the source, until the absorption is so high that it is equal to the rate of supply and there is no further increase in energy density and intensity. Instead of eqn 9.3 we have

$$V \frac{d\mathcal{E}}{dt} + S\bar{\alpha}I = W_1 = S\bar{\alpha}I_1$$

$$\therefore \quad \frac{4V}{c} \frac{dI}{dt} + aI = aI_1$$

$$\therefore \quad \frac{dI}{dt} = \frac{-ac}{4V}I + \frac{ac}{4V}I_1$$

$$\therefore \quad I = A \exp\left[-(ac/4V)t\right] + I_1$$

If we take $t = 0$ when $I = 0$, then $A = -I_1$ and

$$I = I_1\{1 - \exp\left[-(ac/4V)t\right]\}$$

Reverberation time is associated with the idea of how long it takes for a sound to decay to a level where it is no longer noticeable. It is instructive to calculate the corresponding time required for a sound to rise to a noticeable level. From the definition of reverberation time the noticeable level is 60 dB below the initial level, so we should now find the time required for the intensity level to rise from zero to 60 dB below the final level. We do this by writing

$$-60 = 10 \log (I/I_1) = 10 \log \{1 - \exp\left[-(ac/4V)T'\right]\}$$

$$\therefore \quad \log \{1 - \exp\left[-(ac/4V)T'\right]\} = \bar{6}.000$$

$$\therefore \quad 1 - \exp\left[-(ac/4V)T'\right] = 10^{-6}$$

$$\therefore \quad \exp\left[-(ac/4V)T'\right] = 1$$

$$\therefore \quad \text{to more than four significant figures } T' = 0$$

We see that the sound rises to a significant level of intensity almost instantaneously when a sound source starts radiating, but takes a measurable time to decay to an insignificant level after the source stops radiating. The decay is thus more important than the rise. The growth and decay of a sound field are shown graphically in fig. 9.3.

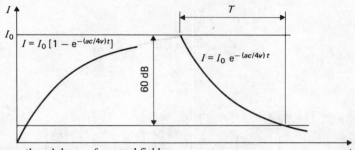

9.3 The growth and decay of a sound field

9.5   Room sound level

The reverberation time is a measure of the rate of decay of sound in a room, and gives a useful indication of the acoustic properties of the room. It is, however, equally important to estimate what the actual sound level in a room will be as a result of a steady source of sound. We again use the fact that the total rate of energy absorption is $I_r S \bar\alpha$. There is also a steady rate of energy generation W_r. With a steady, continuous source, equilibrium will be reached when the rate of energy absorption is equal to the rate of energy generation, that is

$$W_r = I_r S \bar\alpha$$

The total sound power generated by a source in a room is divided into the direct field, which reaches a receiver in the room directly without any reflection, and the reverberant field, which is the diffuse field produced by multiple reflections at the walls, floor, and ceiling. If we imagine the sound travelling outwards from the source, it forms part of the direct field until the first reflection occurs. At that moment a proportion $\bar\alpha$ is lost be absorption, and the remainder, a proportion $(1 - \bar\alpha)$, forms the reverberant field. So

$$W_r = W(1 - \bar\alpha)$$

The whole of the sound power W is radiated as direct energy, but only an amount W_r continues as reverberant energy.

We then have that the intensity of the reverberant field is

$$I_r = W_r / S \bar\alpha$$

$$= W(1 - \bar\alpha)/S\bar\alpha$$

But at a surface in a diffuse field,

$$I_r = p_r^2/4\rho c$$

$$\therefore \quad p_r^2/\rho c = 4W(1 - \bar\alpha)/S\bar\alpha \tag{9.13}$$

For the direct field the free field equation holds

$$p_d^2/\rho c = QW/4\pi r^2 \tag{9.14}$$

where Q is a directivity factor

The total sound pressure is found by adding p_r and p_d according to eqn 2.3

$$p^2 = p_d^2 + p_r^2$$

$$\therefore \quad \frac{p^2}{\rho c} = \frac{QW}{4\pi r^2} + 4W\frac{(1-\bar\alpha)}{S\bar\alpha}$$

$$= W\left[\frac{Q}{4\pi r^2} + 4\frac{(1-\bar\alpha)}{S\bar\alpha}\right]$$

Putting $\rho c = 415$ rayls, and dividing by $W_0 = 10^{-12}$, gives

$$\left(\frac{p}{20.37}\right)^2 \times \frac{1}{10^{-12}} = \frac{W}{10^{-12}}\left[\frac{Q}{4\pi r^2} + 4\frac{(1-\bar\alpha)}{S\bar\alpha}\right]$$

$$\left(\frac{p}{2\times 10^{-5}}\right)^2 \times \left(\frac{1}{1.019}\right)^2 = \frac{W}{10^{-12}}\left[\frac{Q}{4\pi r^2} + 4\frac{(1-\bar\alpha)}{S\bar\alpha}\right]$$

$$\text{or}\quad \left(\frac{p}{p_0}\right)^2 \times \left(\frac{1}{1.019}\right)^2 = \frac{W}{W_0}\left[\frac{Q}{4\pi r^2} + 4\frac{(1-\bar\alpha)}{S\bar\alpha}\right]$$

Taking logs of both sides gives, in decibel notation

$$\text{SPL} - 0.0168 = \text{SWL} + 10\log\left[\frac{Q}{4\pi r^2} + 4\frac{(1-\bar\alpha)}{S\bar\alpha}\right] \tag{9.15}$$

Since it is never necessary to work to more than the nearest decibel, the amount 0.0168 may be ignored. It is more usual to write this equation as

$$\text{SPL} = \text{SWL} + 10\log\left[\frac{Q}{4\pi r^2} + \frac{4}{R}\right] \tag{9.16}$$

where $R = S\alpha/(1-\bar\alpha)$ \hfill (9.17)

R is called the **room constant.**

If S and r^2 are measured in square feet instead of square metres, eqn 9.16 becomes

$$\text{SPL} = \text{SWL} + 10\log\left[\frac{Q}{4\pi r^2} + \frac{4}{R}\right] + 10 \tag{9.18}$$

Eqn 9.16 enables us to calculate the sound pressure level in a room due to a source of noise. Once we know the sound power being generated in a room or the sound power entering it from outside, eqn 9.16 tells us the sound pressure level in the room due to that source. The effect of the absorbent surfaces is contained in the term R. The equation also gives information on how the sound level in a room is affected by the treatment of the walls, again because R is a function of the coefficient of absorption.

The equation is not always straightforward to use because there may be some doubt about finding a suitable value of R, and various semi-empirical approximations are sometimes found in practice. Frequently they are based on the volume of the room instead of the surface area, and this implies that some assumption has been made about the proportions of the room. Again, if r^2 is fairly large compared with $16 \pi/R$, the sound field is almost entirely reverberant and the term $Q/4\pi r^2$ may be ignored.

It can be seen that if $\bar{\alpha}$ is small, $(1 - \bar{\alpha})$ is nearly equal to 1, and R becomes equal to $S\bar{\alpha}$. This is exactly the same approximation as we previously made in deriving the Sabine formula from the Norris—Eyring formula for reverberation time. For strictly logical consistency, if we use the Sabine formula for reverberation time, we should write the sound level in a room as

$$SPL = SWL + 10 \log \left[\frac{Q}{4\pi r^2} + \frac{4}{S\bar{\alpha}} \right] \qquad (9.19)$$

9.6 Transmission between rooms

We are now in a position to combine some of the results already obtained in order to calculate the sound level in one room due to a noise in an adjoining room. We are given the SPL in the source room and wish to find the SPL in the receiving room due to the sound level in the source room.

Let the area of the common partition between the two rooms be S_p, as shown in fig. 9.4. We will use the suffix 1 to refer to the source room and the suffix 2 for

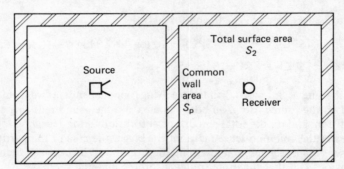

9.4 Transmission between rooms

the receiving room. Then the energy reaching this partition on the source side is intensity times area,

$$W_1 = I_1 S_p$$

The fraction of the incident energy transmitted is the sound transmission coefficient, so

$$W_2 = \tau W_1$$

The sound power entering the receiving room is

$$W_2 = \tau I_1 S_p$$

If the area of the common partition is large, the direct sound energy radiated from it will be much the same at all points in the receiving room. With a large radiating surface forming part of the walls of the enclosure, one can hardly distinguish between the energy leaving that surface as direct radiation and the energy leaving it as a result of reflection, so it is not unreasonable to treat the whole of the energy as being reverberant. Making this assumption, we can simplify our calculation by considering the whole of the entering sound power as going into the reverberant field. Then

$$(SPL)_2 = (SWL)_2 + 10 \log (4/S_2 \bar{\alpha}_2)$$
$$= 10 \log \tau I_1 S_p + 10 \log (4/S_2 \bar{\alpha}_2)$$
$$= 10 \log \tau + 10 \log I_1 + 10 \log 4 + 10 \log (S_p/S_2 \bar{\alpha}) \qquad (9.20)$$

If the sound field in the source room is also diffuse, then the intensity at the wall is related to the sound pressure by eqn 9.2, which may be written

$$10 \log I_1 = (SPL)_1 - 10 \log 4 \qquad (9.21)$$

Also the sound reduction index of the partition is

$$SRI = 10 \log (1/\tau)$$
$$= -10 \log \tau \qquad (9.22)$$

Substituting eqns 9.21 and 9.22 in eqn 9.20 gives

$$(SPL)_2 = -SRI + (SPL)_1 - 10 \log 4 + 10 \log 4 + 10 \log (S_p/S_2 \bar{\alpha})$$
$$\therefore \quad (SPL)_1 - (SPL)_2 = SRI - 10 \log (S_p/S_2 \bar{\alpha}_2) \qquad (9.23)$$

We see that the sound pressure in the receiving room depends, as we would expect, on the SPL in the source room, on both the area and the sound reducing properties of the partition, and on the total amount of absorption in the receiving room. Eqn 9.23 expresses the common-sense fact that the bigger the area of the partition between two rooms, the less sound attenuation there will be between them; and the less obvious fact that the more sound absorbing material there is in the receiving room, the quieter will the intruding sound be.

The presence of sound absorbing material in a room does not provide a very useful means of reducing unwanted sounds, because it reduces the wanted sounds as well. On the other hand, unwanted sound from a plant room or machinery room intruding into an office or bedroom can be reduced very effectively by reducing the level in the source room with sound absorbing material. It may at first seem surprising that a noise problem in one room can be overcome by doing something in a different room, but the essence of good noise control is that attention must be paid to each part of the sound transmission process between the source and the receiver.

Another useful expression is that for the sound power entering or leaving a room through an opening in a wall such as a window. It follows almost immediately from eqn 9.2 by using the fact that power is intensity times area,

$$W = IS$$

$$10 \log (W/10^{-12}) = 10 \log (I/10^{-12}) + 10 \log S$$

$$\therefore \quad \text{SWL} = \text{IL} + 10 \log S$$

$$= \text{SPL} + 10 \log S - 6 \tag{9.24}$$

This is the sound power flowing through an open window from a diffuse field. In a built up environment there is so much reflection from neighbouring buildings that the sound field in the street is more nearly diffuse than directional, and this equation may be used for the sound entering through an open window. Even in rural surroundings, there is often enough reflection from natural objects to allow eqn 9.24 to be used with reasonable accuracy.

We end this theoretical discussion with a note of caution. Before placing too much reliance on any of these equations it is as well to remember the many assumptions that have had to be made in deriving them. The main assumption is that the sound energy is diffused evenly throughout the room. This in turn depends on the reverberation time being sufficiently long for a large number of reflections to take place before all the energy is absorbed. The reflections at the boundaries must occur randomly, which requires that the room has no sound focusing properties, such as can occur in an elliptical room, this is discussed in the next chapter. The proportions of the room must be fairly normal, with no odd shapes or recesses, and it must not be coupled to another room by an opening. An average coefficient of absorption should only be used where there is no excessive variation in the values of the individual coefficients of absorption. Where any of these conditions are not fulfilled, discretion must be used in applying theoretical results to practical problems, but in spite of the assumptions and approximations, useful predictions can be made in most practical cases.

9.7 Enclosure of machines

The same principles that apply to room acoustics hold for any enclosure, including an acoustic housing constructed around a noisy piece of machinery or equipment. The effectiveness of such a housing depends on whether one looks at it from the point of view of a machine operator working inside the housing or from the point of view of people outside the housing.

The walls of the enclosure can never be fully sound absorbent. However little sound is reflected from them, it adds to the sound level within the enclosure. Even if they were perfect absorbers, they could not remove more acoustic energy from the enclosed volume than would leave it by free field transmission in the absence of enclosing walls. Constructing an enclosure around a sound source, where there was none before, will increase the sound level in the immediate vicinity of the source, whatever the properties of the enclosing material. Where an enclosure already exists, however, then the sound level on the inside can be reduced by increasing the absorption of the surfaces.

The increase of acoustic energy inside the enclosure due to reflections from the walls means that there is less energy outside it. From this point of view the property required is sound insulation, as discussed in the chapter on transmission through walls. If in addition the inside faces of the enclosure have sound absorbing properties, the sound level outside will be less simply because the sound level inside is less than in a similar enclosure without absorbent walls.

A most important practical point to remember is that most machines dissipate heat, and any enclosure must be designed so as not to interfere with the ventilation needed to remove the waste heat.

10 Room acoustics

10.1 Reverberation and room acoustics

A room with low absorption has a high reverberation time and is described as being
live. Any source will sound relatively loud in such a room. A room with high absorption has a low reverberation time and is described as being **dead**. The same source
will appear much quieter in a dead room than in a live room. An unfurnished un-
decorated room is live; the high reverberation makes it appear as if the walls are
booming back at one. Putting curtains, carpets, and soft furnishings in the room
deadens it, and all the sounds within it are softened. Many commonly observed
effects can be described in terms of reverberation time, and the explanation points
to the way in which acoustic problems may be overcome.

The essence of reverberant acoustic energy is that it persists for an appreciable
time after the initial source has ceased. It therefore tends to mask any new sounds
which start after the first source has stopped but before the reverberant energy has
been absorbed. The syllables of ordinary speech are a series of almost independent
sources stopping and starting one after another. In certain conditions the reverberant
energy of one syllable can mask the direct energy of the next syllable sufficiently
to interfere with intelligibility. This is one reason why it is easier to understand slow
speech better than fast.

A loud sound will take longer to decay to a level where it does not mask the next
sound than a quieter one which is already nearer to the threshold of masking. This
is why a high reverberation time is more noticeable and more disturbing with loud
speech than with quiet. In an extreme case of a noisy loudspeaker in a public view-
ing gallery the author found he could improve the intelligibility of the speech from
the loudspeaker by partially blocking his ears, thereby artificially raising his private
threshold of masking and reducing the significant decay time.

The reverberation time increases with volume, so that it is higher for a large
auditorium than for a small room. The only way in which it can be reduced to
compensate for the increase in volume is by increasing the total absorption. But
increasing the absorption reduces the intensity which will be achieved from a
source of fixed power output. For any source there will be a maximum volume of
room above which the absorption is invariably too high and the required loudness
cannot be achieved. There is thus a limit to how well even the best orator can
make himself heard in a very large auditorium, and beyond a certain size loudspeakers are unavoidable. In very small rooms, on the other hand, highly absorbent
surfaces are necessary to avoid both the reverberation and the loudness becoming
too high.

A large number of people talking near each other at the same time generate
enough acoustic energy for the reverberant field throughout the room to be higher
than the direct field close to any one speaker. The reverberant field then masks
the local direct field, which is the technical way of saying that the general hubbub
makes ordinary conversation impossible. This is a typical effect of reverberation both
at cocktail parties and in indoor swimming baths, and is another example of the

intelligibility of speech decreasing as reverberation time increases.

In many cases the situation can be vastly improved by providing additional absorption and so reducing the reverberation. But the extra absorption also reduces the loudness and if there is too much absorption the loudness will drop so far that the intelligibility again decreases. At what point the loss of loudness becomes important depends on the distance between the listener and the speaker, and it is useful to remember that no amount of absorption can reduce the loudness below what it would be in the open air.

10.2 Optimum reverberation time

Such general considerations show that it is not possible to give a precise figure for the best reverberation time. The value which gives the best acoustic properties depends on both the size of the room and the use to which it is to be put. For small and medium sized rooms about 0.5 seconds is usually considered to be a suitable value. For public speaking in a large hall, a greater reverberation time is necessary to give more amplification in the larger volume. Suitable values appear to be 0.8 seconds for 15 000 m^3 and 1.5 seconds for 30 000 m^3. Music requires more reverberation than speech, and the optimum values range from 1.0 seconds in small rooms to 2.5 seconds in large churches.

Composers of all periods and all styles have always written music for the particular conditions in which they expected it to be performed, and their style reflects their conscious or subconscious experience of the buildings of their time. It is because of this largely subconscious element that different kinds of music are heard at their best in different surroundings. One would not choose to listen to a piano recital in an opera house, nor to stage a modern opera in a mediaeval cathedral. The dependence between the style of music and the surroundings in which it is heard to best advantage, can be expressed by saying that the optimum reverberation time depends on the kind of music.

In concert halls and theatres the reverberation time depends very largely on the audience. Although the way in which it depends on the number of people present can be reduced by using upholstered seats, it still depends on the ratio of the total volume of the hall to the number of seats. The value of the ratio which should be aimed at in good design is between 6 and 9 m^3 per person. High values of volume to listener ratio give a high reverberation time and a corresponding **fullness** of tone, while low values give a low reverberation time with a **clarity** and **transparency** of tone.

Churches present something of a problem. Although they require a higher reverberation time than other places, especially for organ music, it must not be too high or the preacher will not be able to make himself understood. A low reverberation time also gives a pleasant tone to the singing and so helps to encourage the congregation to sing up.

When music is being recorded, a slightly lower reverberation time is needed than for a live performance, so as to make some allowance for the characteristics of the room in which the music will ultimately be reproduced. Whatever the quality of the recording, the room in which it is reproduced adds its own reverberation characteristics to those which have been recorded, and it does this whatever arrangement of microphones and loudspeakers is used. An exact reproduction of the original sound field can only be achieved with a number of loudspeakers in an anechoic chamber.

10.3 Sound diffusion

The shape of the room and the arrangement of the absorptive surfaces in it also have an influence on the optimum reverberation time. The theoretical calculations assume that the sound is perfectly diffuse, but in fact it never is. Differences in the degree of diffusion are noticeable even when comparing two rooms of the same volume and reverberation time. Rooms with less diffusion should ideally have rather lower reverberation times.

To achieve good diffusion requires the sound to be reflected in completely random directions and also requires approximately the same proportion to be absorbed at each reflection. If the absorptive surfaces are not distributed uniformly around the room, the sound level will tend to be lower near the highly absorptive surfaces than near the less absorbent surfaces. However, the effect is not normally important except in very large rooms with widely differing absorption coefficients. If two surfaces in a room are parallel, repeated reflection can occur between them, and such reflections will not help to diffuse the sound to other parts of the room.

The angle of incidence at which sound energy reaches one wall depends on the angle of reflection at which it left the previous wall and on the angle between the two walls. In any room which has a simple geometrical shape, it is possible to find an angle of incidence at which the sound ray will be reflected from successive walls until it reaches the point at which it was first reflected and reaches it at the original angle of incidence. The sound will then continue to be reflected round and round the same closed circuit until it finally dies away. Such a path is shown in fig. 10.1 But if the closed path is equal to an exact number of wavelengths

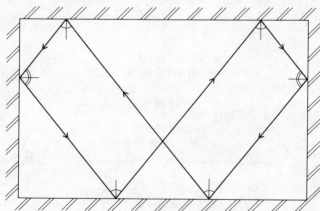

10.1 Resonant reflection

the sound which has just completed a lap will be in phase with, and will reinforce, the sound which is just starting the lap. At the frequency of this wavelength a resonance effect occurs and unexpectedly large amplitudes could be encountered. There will be an intinite number of such closed paths, each corresponding to a different frequency. For the simplest rectangular shape of a room a mathematical expression for the resonant frequencies can be found fairly easily, but it applies to so few practical problems that the equation is not very useful. It is more important to appreciate that any regular arrangement of large flat wall surfaces can give rise to a regular pattern of repeated reflections which set up standing waves at certain discrete frequencies. These are sometimes called the **normal modes** and sometimes the **eigen frequencies**.

Strictly speaking a set of possible resonant frequencies exists for every room, but the more reflecting surfaces there are, and the more random their arrangement relative to each other, the more closely does the sound field approach a theoretically perfect diffuse sound field. For this reason, resonant effects are rarely noticeable in ordinary buildings, but annoying resonances at low frequencies can be set up, sometimes quite unexpedtedly, by an otherwise innocuous source which happens to radiate a frequency equal to a normal mode.

Whilst it is not usually possible to apply the mathematical theory of standing waves to determine the best shape and proportions for a room, the physical ideas lead to a general description of what will give good acoustic properties. The main conclusion is that in order to give a uniform distribution of the possible resonant frequencies and to have as many of them as possible, the major room dimensions should not bear an integral ratio to each other. An irregular asymmetric room will give better diffusion than one of regular geometric proportions. Fig. 10.2 gives some examples of good and bad shapes. The process of diffusing the sound in a random way is improved by an irregular distribution of the absorbent surfaces, but

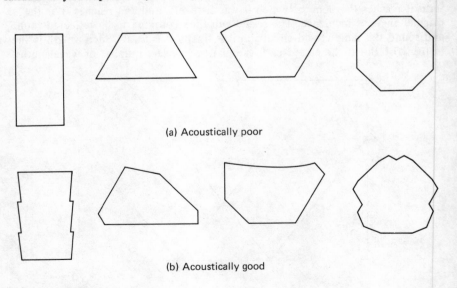

(a) Acoustically poor

(b) Acoustically good

10.2 Auditoria shapes

if this is carried too far it produces the equally undesirable effect of reducing the sound level near the more highly absorbent surfaces and increasing it elsewhere.

A good diffusing surface is a rough one whose roughness is of the same order of magnitude as the wavelength being reflected. A surface which is rough to something small will appear to be smooth to something large. A billiard table, for example, is smooth to the billiard ball but rough to a photon of light. The billiard ball obeys the laws of geometric reflection but the light photons are scattered in all directions giving a visually matt appearance. Random scattering of sound waves requires recesses and projections with dimensions of the order of metres. Good examples of diffusely reflecting surfaces are found in Baroque churches. In church architecture acoustic behaviour is usually an incidental result of aesthetic considerations, but in broadcasting studios deliberate attention is paid to acoustic needs. Cylindrical, spherical, and rectangular surfaces of irregular proportions have all been used to give diffuse reflections in studios.

10.4 Geometric construction methods

When sound is reflected from a relatively rigid surface it obeys, in principle, the same laws of reflection as other wave motions. The reflection of sound is in many ways similar to the reflection of light. In particular the angle of reflection is equal to the angle of incidence, and one can use fairly simple geometric constructions to predict the path which the sound will follow after reflection. This technique ignores the wave nature of the sound (or light) and instead considers rays of sound (or light) travelling along straight lines. It is a perfectly valid technique so long as the dimensions involved are much greater than the wavelength, and in the field of optics very precise instruments are designed by just this means. The wavelength of sound, however, is so large that ray construction methods are not as useful in acoustics as they are in optics. Nevertheless they can be used, particularly for high frequencies and short wavelengths, and where obstructions and reflecting surfaces tend to be large in relation to the wavelength. Providing the results are treated with some reservation they are useful and can give a general indication of the likely distribution of acoustic energy.

Any concave surface will reflect the rays falling on it generally in the direction of the centre of curvature and thereby increase the intensity in a focal region. The conic sections have well known focusing properties, and it is possible to define a geometric focus for them; but any concave surface whose radius of curvature is of the order of the room dimensions is almost certain to give rise to some form of sound focusing. It is better to avoid concave surfaces altogether. Ray tracing is particularly helpful in determining ceiling and wall profiles which avoid unwanted, and perhaps unexpected, sound focusing. Sound focusing is illustrated in fig. 10.3.

An important factor in achieving good acoustics is a short and unobstructed path for the sound from the source to each listener. With a short direct path, the ratio of the direct sound energy to the reflected sound energy is also important, it is desirable to keep this ratio the same for all listeners. It also affects the quality of the sound recorded by a microphone, so that the exact positions and orientations

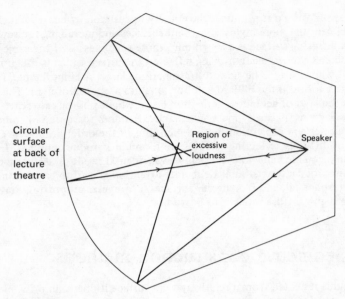

10.3 Sound focusing

of the microphones are of some importance when concerts are being broadcast. It is of course impossible to keep the ratio of direct to reflected sound constant throughout a concert hall or theatre, but various means can be adopted to limit the variation. Raked seating enables direct sound energy to reach the furthest seats. An 8 cm gap between the rays from the source to successive rows of seating has been found to be suitable. The use of balconies and galleries brings more of the audience within closer reach of the source. The plan shape of the auditorium can also be designed to get as many seats as possible as close to the stage as possible. A trapezoidal or fan shape is better than a long rectangle, but a lozenge with the stage in one corner should be avoided because this has undesirable resonance properties.

An echo is noticeable if a significant amount of acoustic energy reaches the listener more than about 40 ms after the original direct sound. In this time sound will travel about 14 m in air, so this is the maximum difference which should be allowed between the length of the direct ray and the length of path travelled by the first reflected ray from the source to the same listening point. A simple geometrical construction can be used to arrive at a shape of boundary surface to meet this requirement. For any point on an ellipse, the sum of the distances from the point to the two foci is constant. So, with the source, S and listening point, R as foci, construct an ellipse such that the length of the ray from focus to focus via any point on the ellipse exceeds the direct distance from focus to focus by 14 m. Then any reflecting surface within the ellipse will not give rise to an echo from the first reflection. Any reflecting surface outside the ellipse must be placed at such an angle that it will not reflect sound into the ellipse. The construction is shown in fig. 10.4.

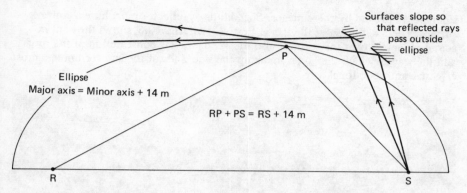

10.4 A construction to avoid echo

10.5 Practical applications

Good acoustic design is as much a matter of what to avoid as of what to put in. Large differences in length between direct and reflected sound paths must be avoided, and the direct sound paths must be as short as possible. Sound focusing by concave surfaces must be avoided. Parallel surfaces should be avoided, and large, flat hard, surfaces should be broken up as much as possible.

Many of these requirements can be met in large auditoria by using shaped reflectors over the stage, and by the use of suspended reflectors over much of the ceiling area. The acoustics of the Royal Albert Hall, fig. 10.5, have been con-

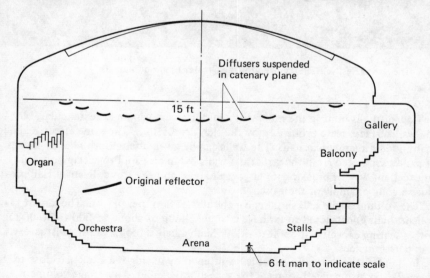

10.5 Longitudinal section through the Royal Albert Hall showing the acoustic treatment

siderably improved by these means. Both kinds of reflector must have a convex surface. The illustration of the Royal Festival Hall, fig. 10.6, shows the convex sound reflectors over the stage. It also shows the zig-zag construction of the roof and the way in which the boxes break up the side walls of the auditorium in a most effective and functional way.

10.6 The Royal Festival Hall (courtesy of the Greater London Council)

Concert halls built in the nineteenth century were generally rectangular, but the wall surfaces were broken up by the rich decorations. The same effect is aimed at in modern rectangular halls. The walls can be given a saw tooth rake by means of panels separated from the structural walls by an air gap. Projecting boxes are an excellent way of breaking up large wall areas, and the Royal Festival Hall provides a good example of their use.

The volume of the hall should give about 7 or 8 m^3 per person. The largest concert halls constructed so far have a total volume of about 25 000 to 30 000 m^3 with a seating capacity of 3500 to 4000. Such a hall is too large for satisfactory solo performances.

In theatres the dramatic effect is heightened by having the audience close to the stage. This is also needed to make the speech intelligible. An average volume is about 3.5 to 4 m^3 per person, and a total volume of about 3000 m^3 with about

800 seats is not uncommon. If the volume per person is greater than this, additional absorption can be provided by making the rear wall highly absorbing. Frequently the rear wall is made slightly convex to improve sound diffusion. Reflecting cylindrical surfaces can also be placed at the sides of the stage.

Opera houses are usually larger than theatres and have slightly more volume per person, about 4 to 4.5 m³. The largest ones have volumes of about 10 000 m³ and such a size makes very great demands on the singer's voice. It is too large for drama with fast dialogue, while the reverberation time is too short for symphonic music. Many of the world's most famous opera houses are horseshoe shaped, which provides the short sound paths required between artistes and audience. The surface is broken up by boxes and deep tiers of seating so that sound focusing does not occur. In both theatres and opera houses a dome ceiling is to be avoided for acoustic reasons in spite of its architectural attractions.

Cinemas should preferably not add anything acoustically to the sound reproduced by the loudspeakers, so they need little reverberation. However, the reverberation time must not be too low or the loudness will vary too much with the distance of the seat from the loudspeaker. A volume of 3 to 4 m³ per person gives a suitable reverberation time without any additional treatment if upholstered seats are used. A saw tooth covering to the walls can be applied as effectively to theatres and cinemas as to concert halls.

Open air music pavilions differ from enclosed rooms, but the same basic principles apply. Their main function is to provide a reflecting surface so as to deflect the sound in the wanted direction. In spite of the need for directional properties, a certain amount of diffusion is needed to avoid reflecting some instruments more than others, and also to provide a satisfactory sound field for the members of the orchestra itself.

Many halls, such as church halls and school assembly halls, are intended to serve a variety of purposes. The best that can be said of many of them is that they are designed for all purposes and suitable for none. When they are full the audience provides enough absorption to make them satisfactory for lectures or talks, solo instrumentalists, and small orchestras, without the addition of any extra absorption specifically for acoustic reasons, but the reverberation time is then too short for a large orchestra or choir. Since most of the absorption is provided by the audience, and since the seats are not usually permanent fixtures, such halls are very resonant when they are empty. If extra absorption is provided for the empty hall, there will be too much when a good audience is present. It is therefore perhaps more important in these halls than elsewhere to ensure good diffusion by means of asymmetric walls, sloping ceilings, projecting beams and columns, and by the use of balconies and galleries; but at the very best a perfect compromise cannot be reached. It is, however, in just these halls that least consideration is given to such matters.

A similar problem is met in gymnasiums. These commonly have a reverberation time of 4 to 6 seconds, which is very high. It can be reduced by providing additional acoustic absorption, particularly at the low frequencies. But if the gymnasium is sometimes wanted for use as a theatre or for concerts, the audience will reduce the reverberation time still further. If the acoustic treatment is satisfactory for concert and theatre performances, then the reverberation time will be too high in the absence of an audience.

Classrooms can more easily be given a suitable reverberation time, although it is rarely done. Common values lie between 1.5 seconds for an empty classroom and 0.9 seconds for a full one. This is higher than is desirable, and can be reduced by lining about 40% of the ceiling with acoustic tiles. It is better to put the tiles round the perimeter of the ceiling and leave the centre as a useful reflecting surface. A slightly sloping back wall improves the quality of the sound at the back of the room. The effect of the floor on the total absorption should not be overlooked. A carpeted floor will give a significantly lower reverberation time than a floor of lino tiles.

In larger lecture rooms the need for good vision from all parts of the room to the lecturer's bench becomes an important design requirement, and almost automatically satisfies the acoustic requirement of short sound paths. A good rake to the seating and a good ground plan, such as a semi-circle or a trapeze, go a long way to meet both the visual and the acoustic needs. The volume of the lecture theatre should be kept as small as possible, and here a sloping ceiling helps. If the ceiling is split into sections of different inclinations it will assist in giving good sound diffusion. Once again saw tooth shaped walls whose surfaces are not parallel to each other are an advantage.

For a church the best compromise between the needs of the organ and the sermon appears to be a reverberation time of about 2 seconds when the church is full and 3 to 3.5 seconds when it is empty. This can often be achieved by means of a wooden ceiling or a suspended plaster ceiling both of which constructions provide absorption at low frequencies. Unsatisfactory acoustic properties have been experienced with concrete ceilings. The danger of providing too much absorbent surface in addition to that on the ceiling is that the reverberation time will be too low when the church is full or even partially full. If extra absorption is needed it could be provided by upholstering the pews, but this is not always a practical proposition, although in some cases the backs of the pews have been given an acoustic lining and then faced with a perforated covering in order to retain the appearance which people have come to expect of church pews. It is difficult to calculate the reverberation time of a church accurately because the shape so often violates the assumptions made in deriving the theoretical formulae. For example, the sound energy density decreases near the ridge of a pitched roof so that the amount of energy absorbed there is less than would be absorbed by the same amount of surface elsewhere. The reduced energy absorption means that the reverberation time is greater than the theory predicts.

11 Vibration and transmission in solids

11.1 Solid vibrations

It sometimes happens that an objectionable noise can be heard in one part of a building, and when it is traced to its source it is found to come from some piece of equipment, such as a pump, in a totally different part of the building. Nothing can be heard between the original source and the room where the complaint is made. Sometimes these can be a very considerable distance apart. If nothing can be heard anywhere between them, then the sound is not being transmitted through the air. The acoustic energy is instead travelling as a vibration through the building structure until it reaches some surface which acts as a good radiator of sound. A common example is a central heating pump which transmits vibrations along the heating pipes to the radiators which then radiate sound as well as heat. Although with this type of transmission the energy can travel quite long distances, noise can also be transmitted between adjacent rooms by vibration of the structure. In this case increasing the sound reduction of the partition between the rooms will not help.

An ordinary building, with all the service pipes and conduits that go into its construction, does not lend itself to a simple analysis of the forced vibrations that can be set up in it. It is in practice unlikely that any reasonable means will be found for increasing the attenuation of any vibration which is being transmitted through it. A much more practical approach is to consider how vibration from the mechanical equipment can be prevented from passing into the supporting structure. If the equipment can be completely isolated, as far as vibration is concerned, then the problem of structure borne noise is completely solved. This can often be achieved with quite a high degree of success.

11.2 Theory of vibration isolation

The simplest case is that of a mass mounted on a spring, as in fig. 11.1. Suppose a force $F = F_0 \cos 2\pi ft$ is applied to the mass, the force acting in the direction of the spring. If the spring deflects by an amount x there will be a spring force sx opposing the applied force; s is the spring constant defined as the ratio of force to extension measured in N m^{-1}. In most cases there will also be some kind of friction or damping opposing the movement of the mass. Usually the damping is small and we can obtain sufficiently reliable results by assuming that the damping is proportional to the velocity. If the damping coefficient is R then the damping force is $R \, dx/dt$ and the total force opposing the applied force is $sx + R dx/dt$.

The equation of motion is

$$F_0 \cos 2\pi ft - \left(sx + R \frac{dx}{dt} \right) = m \frac{d^2x}{dt^2}$$

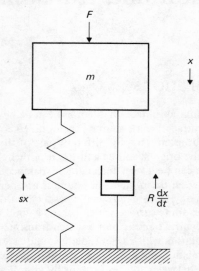

11.1 A simple vibration isolator

or

$$m \frac{d^2x}{dt^2} + R\frac{dx}{dt} + sx = F_0 \cos 2\pi ft \tag{11.1}$$

The steady state displacement x, ignoring transients, will be of the form

$$x = x_0 \cos (2\pi ft + \phi)$$

so that

$$\frac{dx}{dt} = -2\pi fx_0 \sin (2\pi ft + \phi)$$

and

$$\frac{d^2x}{dt^2} = -4\pi^2f^2x_0 \cos (2\pi ft + \phi)$$

Substituting these values in eqn 11.1 gives

$$-m4\pi^2f^2x_0 \cos (2\pi ft + \phi) - R2\pi fx_0 \sin (2\pi ft + \phi) + sx_0 \cos (2\pi ft + \phi) =$$

$$F_0 \cos 2\pi ft$$

To find the values of x_0 and ϕ we consider the particular times for which $t = 0$ and $2\pi ft = \pi/2$. The first gives

$$-m4\pi^2f^2x_0 \cos \phi - 2R\pi fx_0 \sin \phi + sx_0 \cos \phi = F_0 \tag{11.2}$$

while the second gives

$$-4m\pi^2f^2x_0 \sin \phi + 2R\pi fx_0 \cos \phi + sx_0 \sin \phi = 0$$

$$\therefore \quad (s - 4m\pi^2f^2) \sin \phi = -2R\pi f \cos \phi$$

$$\therefore \quad \tan \phi = \frac{-2\pi fR}{-4\pi^2f^2m + s}$$

The expression for tan ϕ enables us to write down sin ϕ and cos ϕ and these can then be substituted in eqn 11.2. We have

$$\sin \phi = \frac{-2\pi fR}{[4\pi^2 f^2 R^2 + (s - 4\pi^2 f^2 m)^2]^{\frac{1}{2}}}$$

$$\cos \phi = \frac{s - 4\pi^2 f^2 m}{[4\pi^2 f^2 R^2 + (s - 4\pi^2 f^2 m)^2]^{\frac{1}{2}}}$$

and hence

$$\frac{-4\pi^2 f^2 m x_0 (s - 4\pi^2 f^2 m)}{[4\pi^2 f^2 R^2 + (s - 4\pi^2 f^2 m)^2]^{\frac{1}{2}}} + \frac{s x_0 (s - 4\pi^2 f^2 m)}{[4\pi^2 f^2 R^2 + (s - 4\pi^2 f^2 m)]^{\frac{1}{2}}}$$

$$+ \frac{4\pi^2 f^2 R^2 x_0}{[4\pi^2 f^2 R^2 + (s - 4\pi^2 f^2 m)^2]^{\frac{1}{2}}} = F_0$$

$$\therefore \quad \frac{(s - 4\pi^2 f^2 m)(s - 4\pi^2 f^2 m) + 4\pi^2 f^2 R^2}{[4\pi^2 f^2 R^2 + (s - 4\pi^2 f^2 m)^2]^{\frac{1}{2}}} x_0 = F_0$$

$$\therefore \quad x_0 = \frac{F_0}{[4\pi^2 f^2 R^2 + (s - 4\pi^2 f^2 m)^2]^{\frac{1}{2}}}$$

We have now got an expression for the vibration of the mass in terms of a maximum displacement amplitude x_0 and an angle ϕ, which gives the phase difference between the applied force and the displacement, when they both vary sinusoidally.

For our immediate purpose we are more interested in the force transmitted F_t through the spring to the foundation. The force transmitted by the spring must be the spring force sx. We can write

$$F_t = sx$$

$$= sx_0 \cos (2\pi ft + \phi)$$

$$= \frac{sF_0}{[4\pi^2 f^2 R^2 + (s - 4\pi^2 f^2 m)^2]^{\frac{1}{2}}} \cos (2\pi ft + \phi)$$

The presence of the spring has changed the force transmitted to the foundation in the ratio

$$T = \frac{F_t}{F_0} = \frac{s}{[4\pi^2 f^2 R^2 + (s - 4\pi^2 f^2 m)^2]^{\frac{1}{2}}} \tag{11.3}$$

and this is the expression which we must examine a little more closely, it is plotted in fig. 11.2. It varies with the frequency, and will reach a maximum when

$$4\pi^2 f^2 R^2 + (s - 4\pi^2 f^2 m)^2 = 0$$

R is usually so small that this critical frequency, which is also called the resonant or natural frequency f_n, is given with sufficient accuracy by

$$s - 4\pi^2 f_n^2 m = 0$$

$$T = \frac{s}{[4\pi^2 f^2 R^2 + (s - 4\pi^2 f^2 m)^2]^{\frac{1}{2}}}$$

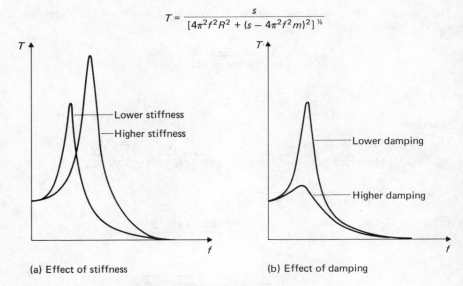

(a) Effect of stiffness   (b) Effect of damping

11.2 The force transmitted by a vibration isolator

or

$$2\pi f_n = \sqrt{(s/m)} \tag{11.4}$$

At this critical frequency the transmitted force is limited only by the damping, and in the absence of any kind of frictional force would become infinite.

At $f = 0$, $T = 1$ and the force transmitted is equal to the applied force for all values of stiffness and damping, as would be expected for a static load.

As f increases from 0 to f_n, $(s - 4\pi^2 f^2 m)$ reduces from s to 0 and the force transmissibility increases from approximately 1 to a very large value.

As f increases above f_n, $(s - 4\pi^2 f^2 m)$ becomes negative, $(s - 4\pi^2 f^2 m)^2$ increases and the **transmissibility** T gets steadily smaller. When $4\pi^2 f^2 m$ becomes equal to $2s$, T in the absence of damping would again become 1. At greater values of f, T continues to get smaller until at very large frequencies

$$T \approx 1/4\pi^2 f^2 m$$

At these large frequencies other more complicated modes of vibration are likely to be set up and the actual value of T may be less than the simple theory suggests.

11.3 Practical application

The important conclusion is that at frequencies greater than $f = \sqrt{2} f_n$, the force transmitted through the flexible mounting is less than the applied force. This is the fundamental principle of all anti-vibration mountings and is the basic means of reducing the direct transmission of sound from a source into the structure.

It is clearly important to choose a form of support which will give a value of f_n sufficiently far below the range of frequencies which it is desired to isolate. One way of reducing f_n is to increase m. This is very often done by mounting heating pumps on a heavy concrete base and supporting the entire base on a resilient layer of cork.

The presence of damping always reduces the force transmitted. The amount of the reduction depends on the relative magnitudes of the damping factor R and the stiffness s. Increasing the stiffness s raises the critical frequency. It also reduces the relative importance of the damping, and will thus increase the force transmitted at all frequencies except in a narrow band around the original critical frequency. The indiscriminate stiffening of a structure is as often as not the wrong solution to a vibration problem; it changes the problem rather than solves it, and may change it for the worse.

Often the stiffness is a constant independent both of the load and of the rate at which the load is applied. In that case the critical frequency can be related to the static deflection. The stiffness or spring rate is given by

$$s = mg/d \tag{11.5}$$

where d is the deflection due to a static load of mg.

Substituting in eqn 11.4 gives

$$2\pi f_n = \sqrt{(mg/md)}$$
$$= \sqrt{(g/d)} \tag{11.6}$$

This relationship is substantially accurate for steel springs. It cannot be applied quite so directly for other types of resilient support for which the deflection does not vary linearly with the load; or which show a hysteresis effect so that they do not return to the original position when the load is removed; or for which the deflection depends to some extent on the rate at which the load is applied. But these effects are usually small enough for eqn 11.6 to give a fair indication of the frequency above which a particular material will give a reasonable amount of isolation. The choice of a suitable isolator is in any case a matter of judgment guided by calculation, rather than of precise selection by numbers, and the static deflection provides a useful guide to the various materials available.

As has just been said, not all materials deflect at a uniform rate as the load on them is increased. The same increase in load may produce a bigger increase in deflection when it is added to an already high load than when it is added to a light load. A non-uniform support of this kind does not have a uniquely defined elasticity, and care is needed in using the correct value for the 'spring rate' when calculating the natural frequency, The **static elasticity** is the total load divided by the total deflection. The **dynamic elasticity** is the increase in load divided by the increase in deflection. If the load—deflection curve is plotted on graph paper, as in fig. 11.3 then the static elasticity at a given load is the slope of the line joining the origin to the point corresponding to that load. The dynamic elasticity is the slope of the tangent to the curve at that point. Only if the graph of load against deflection is a straight line passing through the origin are the static and dynamic values the same. For a very large fluctuation in load it may be more appropriate to use the chord joining the points of minimum and maximum load instead of the tangent at the point of mean load.

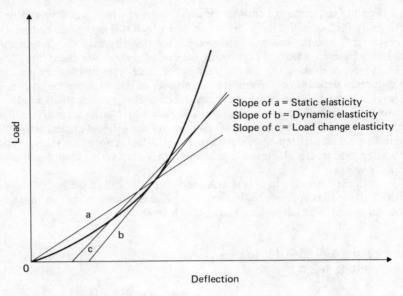

Slope of a = Static elasticity
Slope of b = Dynamic elasticity
Slope of c = Load change elasticity

11.3 Static and dynamic elasticities

If a machine is supported on more than one spring the distribution of the weight amongst them can be found by the ordinary methods of statics, and the individual load on each spring must be used in estimating the critical frequency. If a load is distributed over a relatively large slab of material such as cork or felt, the pressure rather than the weight is used in estimating the static deflection and critical frequency.

Machines are rarely mounted on a single spring, and the possible modes of vibration are more varied and more complex than those allowed for in the simple theory. We have considered displacement in only one dimension, whereas the more detailed study of vibration takes into account both displacement and rotation in three dimensions. It also takes into account a variety of boundary conditions corresponding to different methods of supporting the vibrating body. Any flexibility of the supports themselves can also be considered. The resulting mathematics shows that in general there is more than one critical frequency. However, the basic principles are contained within the simple theory outlined here, and it provides an adequate guide to the practical solution of most of the purely acoustic problems met with in practice. A more detailed analysis of vibration is usually only necessary if the vibration is so large that it becomes a problem in itself rather than forming part of a wider problem of sound transmission.

The commonest forms of anti-vibration mounting are steel springs of all kinds; cork, which is usually used in the form of mats or slabs; felt, also used in mat form; and rubber, which is usually supplied as part of a proprietary mounting. Although rubber is easily deformed, it is not easily compressed. Its apparent compression under load is due to the rubber flowing out sideways and the shortening in one dimension is always accompanied by an increase in another dimension. When using rubber it is therefore important to design the mounting so as to allow plenty of room for the lateral deformation without which a vertical displacement will not be possible.

11.4 Floating floors

The same principles that apply to the resilient mounting of machines can be applied to the construction of the entire floor of a room. A floating floor is simply a thin floor slab supported by a resilient or elastic layer on a much thicker structural slab. The intermediate resilient layer can be a continuous one of fibre glass, felt, cork, or similar material, or it may be made up of individually spaced spring-like supports. Typical structures are shown in fig. 11.4.

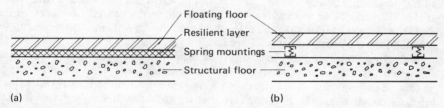

(a)   (b)

11.4 Floating floors

The floating floor and its resilient support form a mass-spring system whose natural frequency depends on the mass of the floating floor and on the stiffness of the resilient support. At frequencies below this natural frequency the floating and structural floors vibrate approximately in phase with each other. The composite floor construction moves more or less as one body and the sound transmission loss is very nearly as predicted by the mass law. Above the natural frequency the two parts of the construction move out of phase with each other and the amount of vibration transmitted to the structural slab decreases as the frequency increases. Since the energy transmitted to the main slab is reduced, the acoustic energy which it in turn radiates to the room below is also reduced.

As with all other vibration transmission problems, the expected improvements will not be achieved if there are flanking paths which allow more energy to be transmitted round the elastic support than is transmitted through it. Careful attention is therefore needed to the constructional details where the floating floor meets the walls of the room.

The vibration of the floating layer can set up acoustic waves in the air between the two layers, and the air waves in turn will set up vibration of the structural layer. Thus the air provides a flanking path round the elastic supports. The airborne waves, like the bending waves in the two solid layers, usually travel in a horizontal direction. If individually spaced supports are used the space between them should be filled with sand or treated in some other way to prevent the transmission of airborne sound through the void.

11.5 Impact noise

In general, the floating floor construction reduces the transmission of all vibrations reaching it, whether they reach it as airborne sound, or as solid-borne vibration from machinery standing directly on the floor. But it also reduces the generation of noise due to impact. The commonest source of impact noise is footsteps, but it can also be caused by anything falling on to a floor. An impact is difficult to measure, but it may be regarded as a very large force acting for a very short time, and the magnitude of the impact is the product of the force and the time.

Any surface can radiate sound as a result of being given a sudden blow. The short sharp blow sets the floor slab into vibration, and the vibrating surfaces radiate sound. The amplitude and frequency of the vibration depend on the magnitude of the impact, on its location relative to the supports of the floor slab, on the material and dimensions of the slab, and on whether the blow happens at a point, or along a line, or whether it is spread over a small but nevertheless finite area. If the blow strikes a floating floor then at frequencies well above the natural frequency of the floor the force transmitted to the main slab is much reduced, and so the sound radiated into the room below is reduced.

A floating floor is not as simple a system as an idealized point mass on an idealized spring, but its action can be analysed in more complex ways. Thus impact noise can be treated as being due to the vibrations which are set up in the floor. With a floating floor construction the vibrations are first set up in the floating layer. If this layer is sufficiently thick and rigid it will itself dissipate much of the impact energy, and so reduce the energy available for transmission to the structural slab. Of the energy reaching the structural slab some is used up in overcoming the internal resistance to vibration of this slab, and only the rest is available for re-radiation as sound into the room below.

Just as the mass supported on a simple spring can increase the force transmitted at frequencies very near the critical frequency, so a floating floor can increase the impact sound levels near its critical frequency. In choosing a suitable construction one must make sure that the natural frequency is below the audible range.

11.6 Structure-borne sound

The only kind of waves that can be transmitted through a fluid are compression waves. A solid on the other hand can be bent and twisted as well as compressed, and so other types of waves are possible in a solid. The type of wave depends on the type of deformation. Apart from compression waves, shear waves, torsion waves, and bending waves can all be transmitted through a solid. In each case the velocity at which the wave travels depends on the density of the solid and on the elastic modulus appropriate to the particular type of deformation.

As the waves travel through the solid, they must do work against the internal friction which resists the deformation. The energy so used is converted into heat and reduces the amplitude of the travelling wave, just as the fluid compression wave is attenuated in passing through the atmosphere. Although this attenuation is often greater in solids than in fluids, it is still not big enough to give a noticeable rise in temperature, nor is it enough to damp the waves so quickly that we could always afford to ignore them.

The mathematical treatment of the many ways in which bars, rods, beams and plates can vibrate is more involved than the relatively simple theory of compression waves in fluids. Nor is it always easy to make suitable simplifying assumptions about the complicated arrangements of solids, which are found in buildings and other engineering structures to enable the mathematical theory to be applied to them. Because of the complexity of both the theory and its application to practical problems, there is no quantitative approach to structure-borne sound as straight-forward as that for airborne sound. Nevertheless the analysis of airborne sound points the way to some general qualitative conclusions about the kind of things which will affect the transmission of sound through structures.

In particular, reflection will occur wherever there is a change either in the direction of propagation, or in the area of the path of propagation, or in the material of propagation. At all such places there will be a reduction in the energy transmitted. Although the actual transmission loss cannot usually be estimated with any degree of confidence, we can say as a general rule that the bigger the change in the transmission path, or the bigger the mis-match between the two parts of the transmission path, the bigger will be the transmission loss at the junction.

In a few cases it is possible to introduce a deliberate mis-match at a junction between two solids in order to reduce the transmission of vibration. It is, for example, good practice to insert felt or other soft material between ventilating ducts and their supports. Various types of plastic inserts can be used with pipe brackets, either between the pipe and the bracket or between the bracket and the supporting wall. Both these cases may be regarded as causing a reflection of energy at a boundary between two materials of widely different properties.

12 Radiation of sound

12.1 Elementary sources

Any surface will radiate sound if it is in contact with a fluid and vibrates at a frequency in the audible range. The layer of fluid immediately in contact with the surface moves with the surface, and the movement of the surface layer is enough to set up the train of compressions and rarefactions which make up the sound wave travelling through the fluid. Usually the fluid we are interested in is air, and the surface is either part of a building structure such as a wall, or the surface of some machine. Sometimes we have to deal with sound transmitted under water, perhaps from the side of a ship. The same general principles, however apply to the interaction of any solid with any fluid.

The simplest source to analyse mathematically is an infinitely small sphere whose surface vibrates in the radial direction, all points on the surface moving in and out together. The particles of air immediately in contact with the surface of the sphere will have a radial displacement equal to that of the sphere. We can apply the equations for spherical waves to find the particle displacement at any distance from the centre of the source, and from the relationship between the particle velocity and acoustic pressure we can also find an expression for the acoustic pressure in terms of the amplitude of the surface velocity of the source. This type of source is known as a **monopole**. Another simple ideal source is a hemisphere, which is exactly like a spherical source except that it radiates only on one side of a plane surface passing through it.

It may seem that such idealized sources are not of much practical value, but real sources can often be analysed as a combination of simple sources. As an example, consider a plane surface vibrating as a whole, the whole surface being displaced in the direction of its own normal but without being distorted. Suppose such a plane is replaced by a large number of hemispherical sources distributed over the whole area of the vibrating surface. If the surface displacements of all the hemispheres are resolved parallel to, and normal to, the surface, all the components parallel to the surface cancel out and only the normal components are left. The resultant displacement of all the distributed sources is thus equal to the actual displacement of the real surface. Mathematically they are identical. At some distance from the surface the total acoustic pressure will be the sum of the acoustic pressures due to each source, and can be found by integrating the contributions of all the simple sources.

In general, a theoretical analysis of any source is possible only if it can be replaced by equivalent arrangement of point sources. In theory at least, the total acoustic pressure at any point can be found by summing the contributions of all the simple sources. In practice the mathematical expressions can only be integrated in a few simple cases and then only by making suitable assumptions. If a more precise solution is needed in a specific case, a numerical integration could be carried out with a computer, providing an appropriate distribution of spherical or hemispherical sources could be established.

The point sources used for this mathematical exercise need not vibrate in phase with each other. For example, when a plate vibrates, sound will be radiated from both the back and front surfaces. If a displacement away from the surface is regarded as positive, then a positive displacement on one side of the plate is accompanied by a negative displacement on the other side. In other words, the two sides of the plate are vibrating out of phase with each other. If the plate is small relative to the wavelength of sound at the frequency being radiated, the sound will bend or refract round the plate. At some distance from the plate the total acoustic pressure will be the sum of the acoustic pressures generated by each side. The two component pressures are not the same, because the two sources are radiating out of phase with each other and in general their distance from the point is slightly different. When dealing with this type of source the concept of a dipole is useful. **A dipole** consists of two simple sources of equal strength which are a very short distance apart and which radiate out of phase with each other. This is shown in fig. 12.1.

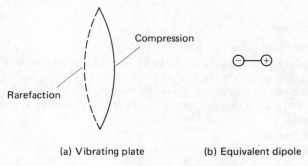

(a) Vibrating plate   (b) Equivalent dipole

12.1 An acoustic dipole

Dipole sources occur in several other cases. An important one is the turbulent flow of fluids. The characteristic of turbulent flow is that portions of the fluid very close to each other have local motions in opposite directions. These small local motions are constantly changing, and the resulting fluctuations are a source of sound. The radiation of sound from turbulent flow behaves as if it came partly from a distribution of dipoles. Similar out of phase motions occur at the edges of a surface which is vibrating in such a way that different sections are moving out of phase with each other. Adjacent areas at the edge then form a line of dipoles along the edge.

Wherever adjacent areas are moving out of phase with each other there is a nodal line between them at which there is no movement of the surface. In general, there are a number of nodal lines which run across the surface in both directions. A detailed analysis shows that in every case the nodal lines meet the edge at right angles and that they cross each other at right angles, as illustrated in fig. 12.2. Where two lines cross four areas meet, two of which vibrate in phase while the other two vibrate in the opposite phase. The sound field which is produced by this kind of vibration is similar to that which would be produced by four simple sources or a pair of dipoles. Such a source is referred to as a **quadrupole**. Examples of quadrupole sources are also found in turbulent flow.

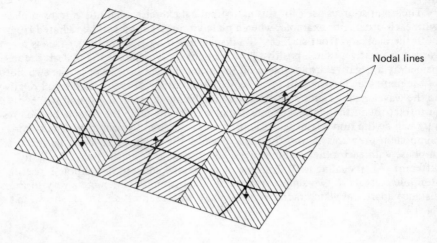

Nodal lines

12.2 A vibrating sheet

As may be expected, a dipole is a less efficient radiator of sound than a monopole, and a quadrupole is even less efficient. If, as is sometimes possible, an actual source of sound can be analysed into some arrangement of monopoles, dipoles, and quadrupoles, then it may be permissible to ignore all except the monopole part of the radiation.

12.2 Surface waves

A relatively thin solid, such as a sheet or plate, can distort by bending of the surface. The distortion can travel across the surface obeying the same laws of wave motion as apply to the compression waves of sound travelling in air. If the amplitude is sufficiently large the waves can actually be seen.

Travelling surface waves can radiate acoustic energy into the surrounding fluid. The frequency of the compression wave in the fluid must be the same as the frequency of the bending wave in the solid, but the wavelengths will be different. In each case the wavelength will be the wave velocity divided by the frequency. Now the wave in the fluid can be resolved in two directions at right angles to each other, one parallel to the surface and one normal to the surface. So long as the surface and the fluid remain in contact everywhere, the component of fluid wavelength parallel to the surface must be equal to the wavelength of the surface bending wave. This then determines the component of fluid wavelength parallel to the surface, as in fig. 12.3. But the resultant wavelength is determined by the frequency and the acoustic velocity in the fluid.

If the velocity of the surface bending wave is equal to the acoustic velocity in the fluid, the wave in the fluid will be propagated parallel to the surface. If the surface velocity is higher than the fluid velocity, the surface wavelength will be less than the fluid wavelength and the fluid wave will be propagated at some angle

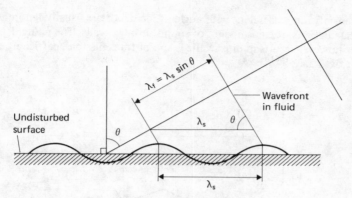

12.3 Wavelengths in a fluid and on a solid surface

θ to the normal to the surface. The angle is given by

$$\lambda_f = \lambda_s \sin \theta$$

or

$$V_f/f = V_s/f \sin \theta$$

If the surface velocity is less than the fluid wave velocity no real value exists for θ and radiation into the fluid is impossible. This means that a bending wave travelling along the surface of a solid can only transmit acoustic energy into the neighbouring fluid if the velocity of the surface wave is greater than the acoustic velocity in the fluid. The velocity of the surface wave generally increases with frequency and no significant amount of acoustic radiation will occur at frequencies below those where the two velocities are equal.

Energy can also be transferred from the fluid to the solid. If a compression wave in the fluid reaches the solid surface, it can set up bending waves at the surface. In this case also the surface wavelength will be the component of the fluid wavelength resolved along the surface. The frequency is of course the same as in the fluid, and so the velocity of the surface wave is determined. This is a forced velocity which does not depend on the properties of the solid in any way.

If one side of a panel is set into forced vibration in this way the other side is bound to fulfil the condition for radiating acoustic energy into the fluid. Thus sound can be transmitted through a panel by the sound setting the entire panel into vibration. Analysis of such transmission leads to results very similar to those which we obtained previously, namely that the transmission loss depends on the surface density and on the angle of incidence.

At any one frequency there is a particular angle of incidence for which the forced velocity is equal to the natural velocity of surface waves in the particular solid. Similarly for any angle of incidence there is one particularly frequency for which the two velocities are equal. For such a combination of frequency and angle of incidence the transmission is very high and the transmission coefficient is very low, this is known as the coincidence effect. Fortunately in practice one is usually dealing with random diffuse sound fields and with random incidence, and variations of transmission coefficient with angle of incidence are usually ignored in practical calculations of transmission of sound through walls. The coincidence effect does however results in dips in the graph of transmission coefficient against frequency, fig. 12.4.

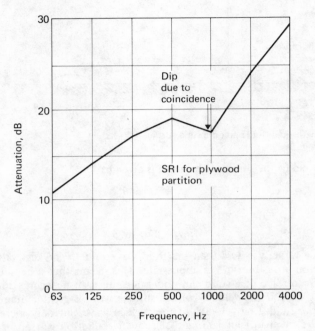

12.4 The coincidence effect

12.3 Vibrating panels

Suppose a wave meets a discontinuity in its path and is reflected back along the same path until it meets another discontinuity where it is again reflected into the original direction of travel. A series of repeated reflections will be set up of exactly the type we considered when discussing transmission past two boundaries in Chapter 5. When equilibrium is reached there will be two waves travelling in opposite directions between the two boundaries. For simplicity we will for the moment consider that they are of equal amplitude. Then if the boundaries are separated by a distance l the two waves can be represented by

$$p_1 = A \cos 2\pi(ft - x/\lambda)$$
$$p_2 = A \cos 2\pi[ft + (x - l)/\lambda]$$

If the distance l is an integral multiple of the wavelength, the second wave becomes

$$p_2 = A \cos \left[2\pi(ft + x/\lambda) - n2\pi\right] = A \cos 2\pi(ft + x/\lambda)$$

and the sum of the two waves is

$$p = p_1 + p_2 = A \cos 2\pi(ft - x/\lambda) + A \cos 2\pi(ft + x/\lambda)$$

$$= 2A \cos 2\pi ft \cos (x/\lambda)$$

This is a **standing wave** and is shown in fig. 12.5. At any time t it is distributed

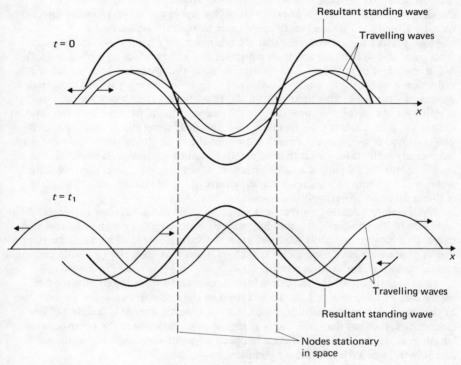

12.5 A standing wave

sinusoidally along the x-direction, and at any point x along the line it varies sinusoidally with time t. But it has zeros or nodes at fixed points given by $x/\lambda = (n + 1)\pi/2$ and anti-nodes fixed in space at $x/\lambda = n\pi$. Unlike the travelling wave, the positions of its nodes and antinodes are independent of t. The crest of the wave does not move in the x direction.

If the amplitudes A and B of the two waves travelling in opposite directions are not equal, then their sum is

$$p = A \cos 2\pi(ft - x/\lambda) + A \cos 2\pi(ft + x/\lambda) + (B - A) \cos 2\pi(ft + x/\lambda)$$

$$= 2A \cos 2\pi ft \cos (2\pi x)/\lambda + (B - A) \cos 2\pi(ft + x/\lambda)$$

This is a standing wave, given by the first term, with a travelling wave, given by the second term, superimposed on it. Thus waves travelling in opposite directions will still set up standing waves, providing their phase relation is correct, even if they differ in amplitude. The effect of the travelling wave superimposed on the standing wave is to blur the nodes. In practical cases of sound radiation, there is no reason why the two amplitudes A and B should be significantly different, and the term containing $(B - A)$ will be negligible, except to explain why sharply defined nodes and anti-nodes cannot always be obtained experimentally.

We have considered the wave to be reflected at two places, and have assumed the directions of incidence and reflection are normal to the reflecting surface. But it is not necessary to impose these restrictions. The same analysis shows that a standing wave will be set up wherever travelling waves can travel in either direction round a closed path whose length is an exact number of wavelengths.

Now just as sound waves are reflected whenever they meet a discontinuity in their path, so the bending waves which travel along the surface of a plate are reflected at the edges of the plate. The reflections obey the usual rule that the angle of reflection is equal to the angle of incidence. The reflected wave will be reflected again when it reaches another point on the edge. It may then happen that a wave is reflected at a number of points round the edge of the plate until it arrives back at its starting point, and is then travelling in its original direction. If this happens it will continue to be reflected round and round the same closed path. A similar wave can equally well travel round the same path but in the opposite direction. If the length of this closed path is an exact number of wavelengths, then the travelling waves will combine to produce a standing wave along the closed path. The plate is then said to be vibrating in a resonant mode.

When a plate vibrates in a resonant mode, in most cases there are many standing waves on it. The nodes of the standing waves form a series of nodal lines at which there is no movement, while adjacent areas on opposite sides of a nodal line vibrate out of phase with each other. This is the condition that gives rise to dipole and quadrupole radiation, and is shown in fig. 12.2.

The exact nature of the radiation from a resonantly vibrating panel depends on the size of the panel and on the direction of the nodal lines relative to the edges of the panel. Although a significant amount of acoustic energy is radiated at low frequencies, most of the radiation is at high frequencies, where the acoustic wavelength in air is less than the distance between adjacent areas and it is no longer accurate to regard them as forming multipoles.

For similar reasons a small panel radiates less low frequency energy than a large one. With small dimensions there is interference between the radiation from the corners, whereas in a large panel the corners are too far apart for interference to occur.

What is a high frequency and what is a low one depends on the size of the panel. A small panel has higher resonant frequencies than a large one. The radiation also depends on the bending stiffness and the surface density of the panel. Thus a small, light, stiff panel will more easily radiate high frequencies, whereas a large, heavy, supple panel will more readily radiate at low frequencies. This may be related to musical instruments by remembering that a violin is played at a higher pitch than a double bass.

It is possible for a panel to have different properties in different directions, and such variations will also affect the radiation from it. If the material is not homogeneous it may have a different modulus of elasticity in different directions; and if it is of non-uniform construction the moment of inertia will be different about different axes. A panel re-inforced by ribs will at low frequencies vibrate as one non-uniform panel, but at high frequencies the areas between the ribs will vibrate independently as separate panels.

12.4 Power transmission

Whenever sound is transmitted to a fluid from the vibration of a solid, the amount of power transmitted from the solid to the fluid depends both on the properties of the fluid and on the amplitude of the vibration at the boundary surface. The amplitude of vibration of the surface in turn depends on the magnitude of the applied force and on the properties of the solid. In very general terms we can say that the displacement amplitude is related to the applied force by the stiffness and resistance of the material. The simplest example is a mass on a spring which we have already looked at in a different context.

The applied force is opposed by the internal friction of the solid, by the elastic forces brought into play by the deformation of the solid, and by the inertia of the vibrating mass. But it is also opposed by the corresponding forces in the air set into motion by the vibrating surface.

The presence of the fluid must be taken into account when calculating the displacement due to a force applied to the solid. This means that the force produced on the solid by the fluid must be included in the equations of motion of the solid. The effect may be represented by introducing the ideas of **radiation resistance** and **radiation reactance**.

The radiation resistance is the equivalent of adding to the frictional forces tending to damp out the vibration. It provides one of the means of removing the energy originally applied to the solid. Some of the applied energy must be used to overcome the purely mechanical resistance to motion. In many cases, such as loudspeakers, some of it is used in overcoming the resistance of an electrical network. Most of the applied energy is converted to heat in these ways, and only a proportion is finally radiated as acoustic energy. This last part is regarded as overcoming the radiation resistance. The advantage of the concept of radiation resistance is that the whole of the applied energy may be treated as having to overcome a series of resistances, namely the electrical, mechanical, and radiation resistances.

The concept of radiation resistance is useful in calculating the acoustic efficiency of a source. The efficiency of a source of sound is defined as the ratio of the energy radiated as sound to the whole of the vibrational energy applied to the source. Providing the various resistances are known, it can be calculated as the ratio of the radiation resistance to the sum of all the resistances. For this calculation it is necessary to express all the resistances in the same units; thus the frictional resistance to motion of a loudspeaker cone can be expressed as the equivalent electrical resistance which would have to be added to a purely electrical network to produce the same effect. The amplifier feeding the loudspeaker 'sees' the total equivalent electrical resistance; it does not 'see' how this is in fact divided between the various forms of electrical, mechanical, and acoustic resistance. Radiation resistance can similarly be expressed in equivalent units and added to other forms of resistance. This idea is illustrated in fig. 12.6.

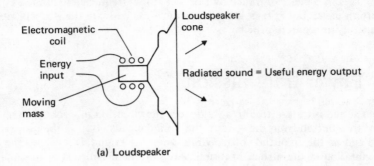

(a) Loudspeaker

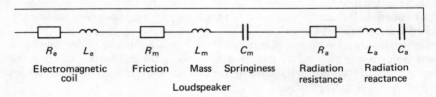

$I^2 R_a$ = Useful energy output

(b) Equivalent electric circuit

12.6 The radiation impedance and equivalent impedance

The radiation reactance is the equivalent of adding to the mass and the elasticity of the solid. It represents a means of storing energy within the system, whereas resistance indicates a means of removing energy from the system. The radiation reactance can be expressed as an equivalent electrical reactance. Electrical reactance, it will be remembered, can be either inductive or capacitive. Inductance is the electrical equivalent of mass or inertia, while capacitance is comparable with elasticity or springiness. If the electrical, mechanical, and radiation reactances are all expressed in the same units they can be added to each other.

Reactances can be combined with resistances, but the addition must be done according to the rules of vector algebra. The total, following the example of electrical engineering, is known as the impedance. The technique of using radiation resistance, radiation reactance, and radiation impedance is useful for calculation purposes, because it brings out a similarity with the mathematics of electrical engineering. Like all analogies, it is useful up to a point, but if pursued too far it can obscure the physical significance of what is actually happening.

Just as electrical impedance is the ratio between a voltage and a current, so radiation impedance is the ratio between force and velocity. As we have said, it is a useful concept when dealing with the coupling between an acoustic source and the load imposed on the source by the surrounding medium.

Other impedances are sometimes used. The **specific acoustic impedance** is the ratio between pressure and particle velocity. It is called specific because it depends only on the type of wave and the medium in which the wave is travelling. For a plane wave the specific acoustic resistance is equal to the product ρc as was shown in Chapter 2. Reference to Chapter 5 will show that it can be used in dealing with transmission from one medium to another and with reflections caused by a change of medium. In fact almost the whole theory of acoustic wave propagation can be developed in terms of specific acoustic impedance without relating this impedance to any other properties of the medium.

The specific acoustic impedance is mathematically analogous to the characteristic impedance which is met in the study of voltage and current waves in electrical transmission lines. The voltage is the analogue of pressure, the current is the analogue of particle velocity. The ratio of voltage to current is an electrical impedance but the specific acoustic impedance has a similar significance and may be thought of as that which prevents pressure differences from producing infinite particle velocities. If there is a difference in phase between pressure and particle velocity, the specific acoustic impedance is expressed as a complex number.

The **acoustic impedance** is the ratio of pressure to volume velocity, and reference to Chapter 6 will show that it can be used in dealing with transmission through one medium past branches and changes of area.

13 Noise sources

13.1 General

Sound will be generated wherever there is a vibrating surface. The surface may be a boundary between a solid and a fluid, or it may be an imaginary surface separating one part of a fluid from another. If energy is provided to make the surface vibrate at a suitable frequency, sound will be generated. The amount of energy required is not large, in fact it is so small that it is usually negligible in comparison with other forms of energy. It is at times instructive to convert the decibel scale to the units conventionally used for measuring other kinds of power and energy. This shows that one watt corresponds to a SWL of 120 dB.

$$SWL = 10 \log (1/10^{-12}) = 10 \log 10^{12} = 120 \text{ dB}$$

The sound pressure level will depend on the distance from the source, the enclosure, and the many other factors discussed in previous chapters, so that the power level alone does not tell us how loud the resulting noise will sound. Nevertheless, without some form of acoustic treatment, a machine generating an SWL of 120 dB would in most cases give rise to an SPL so high that the machine would be considered as being almost unacceptably noisy. Even one milliwatt creates an SWL of 90 dB, and is likely to lead to an SPL which is still regarded as very noisy.

Nevertheless the unwanted noise transmitted from a machine stems from the same things which reduce its mechanical efficiency. By far the greatest part of the energy lost as a result of the inefficiency appears as heat, but the minute amount which appears as acoustic energy is enough to cause all the difficulties of noise abatement of modern life. Anything that improves mechanical efficiency is likely to produce some reduction of noise, although it may take a large change in the efficiency to produce a small change in the sound level.

13.2 Friction

It is then not surprising to find that friction is a common and prolific source of noise. It is most noticeable in bearings and gears. Although there are other sources of noise in a bearing, friction is the most important if a bearing is poorly lubricated or overloaded, and the oil film is inadequate for the load. So long as oil is present it will act as a lubricant, but the film will break down intermittently and give rise to a stick and slip motion which generates sound in exactly the same way as a violin bow moving over the violin string. Overloading of a dry bearing or 'sealed for life' type of bearing has a similar effect. Overloading may be produced by letting a bearing run eccentrically. Even roller bearings do not entirely eliminate friction. The rolling elements always slide a certain amount in their cages, and the noise from this sliding is increased if there is any dirt in the bearing.

Friction is also the most important source of noise in gears, where it appears as the friction between the mating surfaces of the teeth. The point, or line, of contact between two meshing teeth moves over the surface of the teeth as the gears rotate and the teeth slide over each other. As the contact passes the pitch line, the direction of sliding changes relative to the tooth surface and therefore the frictional force also changes direction. This change of direction gives rise to a sound at the frequency of tooth contact. Since the sliding movement is an essential part of the action of the gears, the noise generated cannot be reduced by improving the accuracy of the tooth shape, but just as with bearings, it can be reduced by improving the efficiency of the lubrication and by avoiding overloading.

Friction plays its part in the noise of railway trains where there is rubbing between the sections of the car couplings. In corridor trains there is friction between the end frames where the cars abut.

Any movement of pipework causes friction between the pipes and their supports. This can be the source of unpleasant noises in heating systems, where the expansion and contraction which occur every time the controls switch the heating on or off can produce creaking noises at unexpected times. The movement is of the stick and slip type. The remedy is the use of properly designed and installed supports which take the load imposed by the pipes while allowing sufficient freedom for the anticipated movement between the pipes and the building structure.

In general, lightly loaded slow moving components are quieter than heavily loaded high speed ones. For example, centrifugal fans can be designed to have a slow or fast shaft speed. From the point of view of frictional noise the slower speed is better; but this design requires forward curving blades which introduce aerodynamic noise as discussed in section 13.5.

13.3 Fluctuating loads

We have mentioned that the frictional force between gear teeth changes direction as the teeth pass the pitch line. Whenever there is a change in the direction or magnitude of the load applied to a component there is a change in the stress which is accompanied by a change in the strain. The change in the strain implies that there is some movement, and if the changes occur at a suitable frequency some of the rapidly changing strain energy can be converted into acoustic energy. This condition applies to almost all the ordinary forces acting in a gear train where the loads and strains change continuously relative to the teeth. As would be expected, the noise from this source is greater in highly stressed gears.

Similar fluctuations occur in other forms of power transmission. The well known universal or Hooke's joint does not transmit motion uniformly; although one complete revolution of the input shaft gives one revolution of the output shaft, the relative velocity of the two shafts varies during the course of that revolution. The corresponding variations of stress and strain make the joint a noise source.

Again, a reciprocating engine produces an essentially alternating torque. Although it is provided with a flywheel to even out the transmitted torque the crankshaft torque remains an alternating one which will generate noise at a frequency related to engine speed.

If a wheel rolls along the ground the point of contact moves around the circumference of the wheel as the load is transferred from point to point of the rim; this kind of fluctuation is the cause of tyre-tread noise. As each element of the tread makes contact with the road it compresses, and decompresses again as it leaves the road surface. The resulting stress fluctuation in the tread produces the tyre-tread noise. The same thing happens of course with railway wheels, where any irregularity of the wheels or rails adds to the noise.

Fluctuating stresses cannot be avoided, but their acoustic effect can sometimes be reduced by generous dimensioning.

13.4 Impact

The load changes that occur in ball and roller bearings are slightly different. Here the noise is largely due to the deformation that occurs as the load is transferred rapidly from one ball or roller to the next. No two of the rolling elements are ever exactly the same size, so there is a slight deformation as the load moves from one element to another, and a slight movement of the shaft. These small displacements are enough to produce impact loads on the balls or rollers and the cages. Ball and roller bearings thus generate impact noise which can only be reduced by manufacturing to very close dimensional tolerances.

Impact noise also originates from gears, fig. 13.1 shows how in some gears the

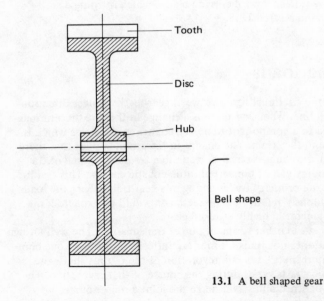

13.1 A bell shaped gear

tooth, disc, and hub form a bell shape which will ring just like a bell as a result of any impact blows as the teeth mesh with one another. Accuracy in the design and manufacture of the tooth form is obviously required to give smooth, impact-free meshing. The shafts must also be sufficiently stiff to prevent any distortion which would interfere with the correct meshing of the teeth. Poor meshing can also be caused by eccentricity due to excessive bearing clearances.

The instantaneous gear ratio depends on the centre distance between gear wheel and pinion, and any error in centering the gears will make the ratio fluctuate about the mean value, just as the transmission ratio of the Hooke's joint varies about the mean value of one. Any such fluctuation produces acceleration forces and impacts which radiate noise.

If any oil, which for this purpose is incompressible, remains in the tooth gap there will be an impact when the meshing tooth strikes it; one can regard the oil as causing a deformation of the tooth shape. On low speed gears there is more time for the oil to drain out of the root clearance before the teeth mesh, but on high speed gears it is particularly important for the oil flow to be designed so that the root clearance does not remain filled with oil.

Sometimes an attempt is made to reduce gear noise by separating the rim of the gear wheel from the disc and hub with a compliant material, but the extra resilience introduced can change the centre distance enough to prevent correct meshing and so add to the noise generated instead of reducing it.

Other sources of impact noise may or may not be obvious. Electrical contacts open and close with an impact which causes noise in switches, relays, and thermostats. Any loose bits of equipment which can be disturbed or dislodged by nearby machinery will generate impact noise. Even crockery in a sideboard can be a source of this kind of noise if the floor is so badly supported that it is set into vibration every time someone walks on it.

Any out of balance forces in rotating machinery can set up corresponding impacts at a frequency related to engine speed. In a petrol or diesel engine the sudden peaks of cylinder pressure distort the engine structure with equal suddenness and generate impact noise.

Moving vehicles in particular are subject to impact noise, which can come from the joints in road surfaces or between lengths of rail, from rain or hail striking the vehicle, or from the sprung weight bumping against stops in the suspension system. The noise due to weather can be controlled to some extent by incorporating damping material in the body to reduce the amount of vibration set up by the impact of rain and hail.

Less obvious forms of impact occur in hydraulic machinery wherever there is a high shock loss in the flow. Thus pumps and fans for handling dust, slurry, or solids, are usually centrifugal machines with fairly widely spaced radial blades, designed to present as little obstruction as possible to the flow of the solid matter. Unfortunately this means that the entry angle to the impeller is the wrong value for good fluid flow and there is therefore a shock loss at the impeller eye. Apart from reducing efficiency, the shock loss creates noise.

Impact noise can often be reduced by good design. It should hardly be necessary to say that all rotating parts must be properly balanced. Adequate stiffness must be provided to prevent unwanted distortion and excessive acceleration. It is good practice to design machine components so that their deflection under the loads imposed during service is substantially less than the specified manufacturing tolerances. Lubrication systems must be designed to allow the free flow of lubricant and to avoid any impact of slugs of liquid.

But whatever the design, noise reduction can also be achieved by accurate machining and careful assembly. A good design can be spoiled by carelessness in manufacture. Conversely, care in manufacturing can mitigate some of the acoustic weaknesses of poor design.

13.5 Hydraulic noise

Few attempts at classifying ever succeed completely, because the things being classified have a habit of overlapping the classifications used. In describing various sources of impact noise we have mentioned noise due to oil in gears and to wrongly shaped pump or fan impellers. Both these could equally well be described as hydraulic noise. The movement of fluid, whether gas or liquid, can generate noise in many places where it may not always be suspected. Gear-teeth, for example, pump air as the teeth mesh and if the clearances between the wheels and the casing are inadequate the unwanted air movement can be a source of noise. A similar air noise can be generated by the rotor of an electrical machine. A more obvious source of noise in gears is splash lubrication. If this is used at high speed it will produce the splashing noise from which it gets its name. Gear-box housings should in general be designed to allow for the free flow of lubricant without excessive splashing or sudden changes in velocity.

Hydraulic noise is the major sound source in fans. It is caused largely by boundary layer separation as the air or other gas flows over the blades. The point of separation varies, and the noise is caused by the formation of fluctuating eddies.

Another source of fan noise is the shedding of vortices at the trailing edges of the blades, fig. 13.2. Any trailing edge which has a finite thickness sheds vortices alternately from opposite sides. All real, as opposed to mathematically ideal, edges must have a finite thickness, however small, so all the trailing edges in practice shed vortices where the flow leaves the edge. The frequency of the noise due to vortex shedding depends on the profile of the blade and its velocity through the air. The velocity always varies along the length of the blade, and the profile usually does, and so the noise generated covers a very broad frequency range. If the blades are too close together it is possible that the eddy formed by one blade does not have time to get clear before the next blade reaches the same point in space, and so the eddy from one blade is struck by the next blade of the impeller. Such interference between a blade and the flow pattern from the previous blade inevitably adds to the noise generated.

In centrifugal fans it is also possible for vortices to form at the leading edges of the blades if the blades form the wrong entrance angle for the air flow. Centrifugal fans with forward curved blades tend to give more aerodynamic noise than those with backward curved blades because the velocity of the air relative to the blades is greater and therefore the turbulence is greater. As has been mentioned in discussing friction, this is offset by the lower shaft speed of the forward curved fan. Both types will be quieter if the blades have an aerofoil shape. In both types a properly designed outlet scroll or volute reduces the shock loses on exit and therefore the sound level.

Axial flow fans tend to operate at higher peripheral speeds than centrifugal fans. They are noisiest when operating near their stall condition. It may be that the pressure

13.2 Vortex shedding from the edge of a rotating blade (courtesy R. Pacifico, Southampton University)

flow characteristic does not show a well marked stall point, but nevertheless the sound generated will be greatest when the fan is operated at conditions where one could expect it to be nearly stalled. Like most machines, a fan will be quietest when operated near its point of maximum efficiency.

An aircraft propeller can of course be regarded as a special case of a fan, and it generates noise in much the same way. Frequently the propeller noise is greater than the engine noise. The vortex noise is particularly important. There is also a rotation noise because the pressure over the propeller disc varies as the blades rotate.

Vortices are also formed by the velocity gradients which occur in any region where two streams meet and mix. The formation of the vortices is generally a random process which gives rise to a noise with a broad frequency band. The higher the relative velocity gradient of the two streams when they meet, the higher will be the noise generated. This is the main source of noise from aircraft jet engines, where the high velocity jet mixes turbulently with the surrounding air. Experiments have shown that the noise is much the same from a cold jet as from a hot one.

One might expect that a jet which produces its thrust from a large mass of gas at low velocity would be quieter than a jet which uses a low mass at high velocity. But it has been found that bypass arrangements which work by surrounding a high velocity central jet with a low velocity outer layer are not significantly quieter than the high velocity core on its own. Nevertheless there are designs of bypass engine which have achieved a significant reduction in noise levels, fig. 13.3. Some reduction in sound power has been achieved by using a shaped nozzle which has a larger perimeter for the same orifice area. By using techniques such as these the noise from the jet engines can be reduced by 6 to 9 dB.

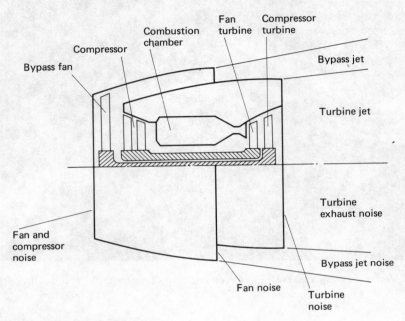

13.3 Noise sources in a bypass engine

When a jet aircraft is in flight the movement of the engine relative to the body of the air produces a ram effect which increases the power output and changes the directivity pattern so that more sound as radiated in the direction of motion. At the same time the movement of the jet relative to the air decreases with aircraft speed, and the jet noise decreases a little.

The sound from a jet is not radiated from the jet nozzle. Measurements indicate that the low frequency noise appears to come from a region five to twenty diameters downstream from the nozzle, while the high frequency noise comes from a region one to nine diameters downstream. One needs to be fifty to a hundred metres away before one can regard the jet as approximating to a point source.

Most aircraft noise is hydraulic noise associated with the propeller, engine exhaust and jet noise, and combustion noise from the engine. Other sources of aircraft noise are the vibration of the engine shell, the gears used in the transmission of power to the propellers, and the vibration of the fuselage skin which is caused by pressure fluctuations set up by the turbulent boundary layer.

A region of turbulent fluid flow is itself a source of sound. Turbulent flow is characterised by rapid fluctuations of the velocity at each point in space, both in magnitude and direction. The fluctuations are local and do not affect the average velocity of flow. The corresponding fluctuations in the flux of momentum across surfaces fixed in space give rise to a distribution of acoustic quadrupoles. Where a turbulent flow meets a solid boundary the same fluctuating forces produce a field of dipoles which form and reform at random. The noise generated by turbulence increases with velocity. It is an important source of noise in ventilation systems and can be reduced by keeping velocities below about five metres per second.

Turbulence, vortex formation and shedding, and boundary layer separation are all sources of unwanted noise and never occur completely independently of each other. They are associated with bends, obstructions to flow, and sharp edges. Boundary layer separation gives a random distribution of frequencies and the amplitude of the sound is approximately proportional to the velocity. The unsteady flow at a fixed sharp edge on the other hand tends to produce discrete frequencies as does the shedding of vortices at such an edge. All are well illustrated by the noise generated at ventilation grilles. The purpose of a grille is to direct the air into the room in a particular manner; it is required to change the flow pattern in the duct into the different flow pattern in the room. There must be some obstruction to the flow to produce the change. If the grille is to produce a wide spread of air into the room, it must produce a bigger change in the flow pattern, or present a bigger obstruction, and indeed a wide spread does give more noise. The noise also increases with flow velocity for all the reasons outlined above. Because the random dipoles and quadrupoles are spread over the entire grille, the noise also depends on the shape and area of the grille.

What are in effect obstructions to air flow are also provided by a moving vehicle. The relative velocities are much higher than would be accepted in any ventilation system, but the general nature of the sound source is exactly the same. These sound sources are particularly noticeable with a motor car. The slipstream moving over the body of the vehicle is subject to boundary layer separation and vortex shedding. An open front window allows the slip stream to impinge on the rear edge of the window frame. The effect can be reduced by an anti-draught device which deflects the slipstream away from the rear of the window.

Flow noise is generated in the inlet and outlet systems of reciprocating engines, and the exhaust noise can often swamp other engine noises. The exhaust is a pulsating flow, and there will be a fundamental frequency related to engine speed. In a multi-cylinder engine the fundamental frequency from each cylinder is out of phase with that from the other cylinders, but some of the harmonics are in phase, so it is possible that the most troublesome frequency will not be the fundamental but one or more of the harmonics. Which harmonics are in phase will depend on the number of cylinders. Reciprocating engines include not only petrol and diesel engines, but air compressors and refrigeration compressors. Reactive silencers are used with all these machines. In addition to reducing the sound transmission as described in Chapter 7, they help to convert the pulsating flow into a steadier flow.

Valves and elbows occur in any piping system and present an obstruction which increases turbulence and noise. They can be listed in the order of increasing noise: a straight length of pipe, an open gate valve, an elbow, an open globe valve, an orifice, a partially open globe valve (a gate valve should not be used for regulating the flow and so is not usually used in a partially open position).

If a valve in a pipe line carrying liquid is shut very quickly, the flow of liquid will be stopped equally quickly. The sudden drop in velocity occurs first at the valve, because all liquids have some compressibility the rest of the liquid can momentarily continue to flow. The liquid that has already stopped is compressed and the pressure at the valve rises. The changes in pressure and velocity are transmitted through the liquid in the pipe at the speed of sound, and can be analysed in just the same way as the transmission of sound through a duct. Reflections occur in the way described in Chapter 7 and dangerously high pressures can sometimes build up. The phenomenon is known as water hammer, although it is by no means restricted to water. It can happen in any hydraulic system, and even if the system is so designed that the pressures produced are not dangerous, it is inevitably accompanied by noise. Since a globe valve gives a more uniform rate of shut off than a gate valve it is less likely to cause noise by water hammer. With a gate valve most of the change in flow takes place over a very small part of the valve movement and the rate of change is correspondingly high. On some installations surge tanks are used to control and reduce the effects of water hammer. It is also possible to use air bottles for the same purpose, but obviously they are only effective so long as they continue to contain air and are not allowed to fill with liquid.

Pressure changes occur in every liquid where there are velocity changes; if the velocity increases then the pressure falls. If the pressure is reduced enough then the liquid will begin to boil and bubbles of vapour will form. This can happen in very small regions where the local velocity is very high. The continued formation and collapse of vapour bubbles is known as cavitation. It has probably been studied mainly in connection with ships' propellers, but it also happens at the inlet of pumps and in any hydraulic equipment where very high local velocities occur. The noise of the collapsing bubbles is not of equal intensity at all frequencies, but tends to be louder with increasing frequency.

A similar noise may be generated in the evaporator of a refrigerator where a mixture of gas and liquid leaves the capillary, but it is usually masked by other noises from the refrigerator or else its transmission is substantially reduced by the insulated cabinet in which the evaporator is housed.

Most of the techniques for reducing hydraulic noise rely on reducing the velocity and turbulence of flow. This may seem very simple, but it is very effective although not always easy to achieve. Generous clearances should be used, and all bends reduced to the minimum. Where bends cannot be avoided they should be of large radius. If bends of small radius must be used they should incorporate splitter vanes, as shown in fig. 13.4. Obstructions with sharp edges should be avoided. Good aerodynamic design and the use of aerofoil shapes wherever possible help to reduce shock losses and the noise associated with them.

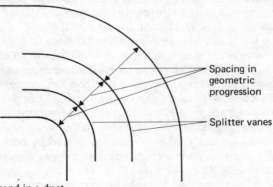

13.4 Splitters in a bend in a duct

It is very difficult to reduce sound radiation from pipework. Although a flexible connection may almost completely stop the transmission of vibration along the material of the pipe itself, it is not completely effective because sound continues to be transmitted through the fluid. Nevertheless a flexible connection does to some extent act as a silencer and the change in material may be enough to set up reflections which reduce the amount of transmission. It has been found that the best material to use is a length of reinforced rubber hose, and reductions varying from 1 to 10 dB per metre have been reported.

Large diameter ducts are often lagged with acoustic material. This acts in two ways. It is a damping material which reduces the vibration of the pipe wall, and it provides an impervious outer layer which is mounted flexibly on the pipe, so that the outer layer vibrates less than the pipe and therefore radiates less sound.

Combustion noise is in some ways similar to hydraulic noise. It consists of random bursts of frequencies up to about 300 Hz. Uneven combustion gives a broad band noise. The use of an afterburner on a jet engine increases the sound level by 5 to 8 dB.

13.6 Resonance

Some resonance effects have been discussed in previous chapters, particularly in connection with vibration isolation and Helmholtz resonators. Almost all equipment has some natural frequency of vibration; this is the frequency at which it will continue to vibrate if left undisturbed after being given a sharp blow. There will also be some frequencies associated with the operation of the equipment, such as the rotational speed of any shafts or wheels, or the frequency of reversal of reciprocating components. If the operational frequencies are close to any of the natural frequencies resonance occurs and all vibrations are greatly magnified. The danger of physical damage from resonance is well known, but even where the amplitude of vibration is not big enough to cause damage it will contribute significantly to the radiation of acoustic energy. All machinery should therefore be designed to have natural frequencies well away from any operational frequencies.

The possible operational frequencies are not always immediately obvious. We have mentioned that the frictional force on a gear-tooth changes directions as the point of contact passes the pitch line. There is therefore a possibility of resonance at the frequency of tooth contact, which is the product of shaft speed and number of teeth.

In ball and roller bearings resonance can occur not only at the frequency of the shaft, but at that of the cage and that of the balls or rollers. There are always small irregularities in the raceways, and two further frequencies arise from these irregularities. The first is given by the difference between the cage and shaft speeds multiplied by the number of rolling elements, and the second by the cage speed multiplied by the number of elements. Thus in a rolling bearing there are five operational frequencies which can give rise to resonance with other machine components. The five frequencies and their harmonics are closely spaced together at low frequencies. Their amplitude will be increased by any wear or damage to the bearing.

Fans can create a resonance effect at the frequency at which the blades pass any point fixed in space. If the number of rotating blades is an integral multiple or sub-multiple of the number of fixed vanes then a lot of noise will be generated. Whatever the two numbers, their product gives a resonance frequency, but the intensity of such product frequencies is usually low. The intensity and harmonic content of fan blade noise depends on the shape of the impulse which the blades give to the air. The momentum or energy transfer from fan to air can be analysed as a Fourier series. The steady component produces the air flow and the fluctuating components produce noise. If the number of blades of an axial fan is increased the number of audible harmonics is reduced but the intensity of each harmonic is unaltered; the overall sound level is therefore reduced.

Fans form an auxiliary part of many machines, and unexpected resonances can occur if the housings and ducts are not well designed and rigidly constructed. Bearings also are not always supported as rigidly as the designer intended them to be and their flexibility has a very great influence on the actual critical speed of the machine concerned.

There is a vast literature on the vibration of rotating shafts, but all that it is necessary to say here is that all rotating parts must be properly balanced both statically and dynamically. Even when balanced, every shaft has one or more critical speeds, and if it rotates at a speed near one of the critical speeds it will be noisy. The effect of the balancing is to reduce the out of balance forces which arise from the design and manufacture and which would set up vibration. A rotating shaft can vibrate in many ways. One is known as whirling and is due to the rotation of the centre of the journal about the centre of the bearing. Thus any eccentricity of the bearing increases the vibration and noise.

Resonance at the natural frequency of one of the mechanical components may be set up by the flow of fluid in a hydraulic system. Any components coming in contact with the fluid should be designed so that their natural frequency is outside the audible range.

There is another possible source of resonance in internal combustion engines between the natural frequency of the engine block and the firing frequency.

In an automobile an open window not only generates noise as already described, but makes the vehicle act as a Helmholtz resonator. The resonance of the body cavity, even with the window shut, can amplify noises originating elsewhere on the vehicle.

Vehicle road noise is associated with the natural frequency of vibration of the tyre casing. The movement sets up a natural vibration in the casing which radiates sound at a frequency of 100 to 200 Hz at all road speeds. The vibration is transmitted through the suspension system to the body of the car where it may be amplified. Another source that radiates at its natural frequency independently of vehicle speed is the brake drum. Brake squeal is due to vibration of the brake drums at frequencies between 10 000 and 15 000 Hz.

Tyre thump on the other hand is associated with an operating repetition frequency, namely the wheel speed in revolutions per second. The harmonics of this frequency set up resonant vibrations of components of the tyre, and the tyre thump is due to the beat between different harmonics. Tyre squeal is due to the vibration of the elements of the tyre tread.

Combustion noise can also be amplified by resonance between the flame and the acoustic source. An increase in pressure increases the combustion rate, so the fluctuating compression associated with the sound wave can produce a fluctuating combustion rate. It is then possible that energy may be converted from the thermal to the acoustic form at just the right moments to increase the amplitude of the fluctuations.

Theoretically, the simplest way of dealing with a resonant noise source is to change the dimensions of the offending component so as to change its natural frequency. It may only be necessary to change the thickness or some other relatively unimportant dimension. If noise is the only consideration, it may be acceptable to let resonance occur but at a higher frequency in the supersonic region.

Alternatively, if the vibration cannot be completely prevented, then its amplitude may be reduced by using adequately rigid supports for bearings and housings. As has already been said, a load bearing component should be rigid enough to make its deflection under load one order of magnitude less than the tolerances to which it is manufactured.

The amplitude of vibration can also be reduced by increasing the friction, which can be done by using suitable damping materials. The Babbitt metal used in bearings is one such high damping material. Rolling bearings are damped if they are a good fit in their housing. Damping can also be achieved by using softer materials such as plastics for load bearing components; plastic pinions in gear trains are quite common. Where damping cannot be achieved by a change of material it may still be achieved by lining the part with a damping coating. This technique is used whenever a sheet metal casing, or housing, could be set in vibration and is commonly used in motor cars. Large ducts for ventilation and other purposes are also treated in this way. The use of anti-vibration mountings, with a natural frequency well below the operating frequency, to reduce the amplitude of vibration has already been discussed in Chapter 11.

13.7 Electric noise

The hum from electric motors, transformers, and chokes is well known to most people, and can contribute to the noise from any machine which incorporates electric motors or controls.

Transformers are wound on cores built up of laminations, and the action of the transformer depends on the variation of the magnetic field in these laminations. Because of the phenomenon known as magnetostriction the laminations change their length as the magnetic field changes. The total change is only a few millionths of the overall dimension, but it is enough to be a major cause of transformer hum. The effect is independent of the direction of magnetization and so a complete cycle of dimensional change takes place for every half cycle of the alternating electric supply. The fundamental acoustic frequency is thus twice the line frequency. If there are harmonics in the electric supply or load, they will produce corresponding acoustic harmonics, but the acoustic harmonics also depend on the dimensions of the transformer. A small transformer will radiate more small wavelength high frequency harmonics, whereas a large transformer will radiate more low frequency harmonics. The tank containing the transformer is a poor radiator of sound at wavelengths greater than its own dimensions.

The amount of noise increases with the magnetic flux density. The flux density is determined by the design of the transformer and does not vary significantly with load, so the sound power level is independent of the load. It is in fact part of the no load loss. Transverse vibrations also contribute to the noise radiation, and can be about as great as the magnetostriction effect. Even if the magnetostriction were completely removed, the transverse vibrations would remain and the sound level would not be reduced by more than about 3 dB.

In practice it is not possible to use materials with a low magnetostriction, nor can the flux density be reduced without increasing the size and cost of the transformer. Improved clamping of the laminations does not reduce the sound power very much. A more practical approach to reducing transformer noise is to reduce the transmission between the core where the noise originates and the tank from which it is radiated. A resilient barrier in the oil between the core and the tank would be a possibility but is not used commercially. Other possibilities are to replace the conventional oil cooling with spray cooling, evaporation cooling, or gas cooling.

Chokes or inductances are very similar to transformers except that they have only one electrical winding. Acoustically this makes no difference. The chokes used to control fluorescent lamps carry a flux which is far from sinusoidal and therefore radiates more harmonics, which are particularly noticeable at about 2000 Hz. The lamp housing can act as a sounding board in the same kind of way as the transformer tank. These chokes are usually provided with an air gap whose purpose is to maintain a fairly constant impedance over a wide range of current. A fluctuating magnetic force is set up across the gap and tends to cause contraction while the magnetostriction is causing extension. The net amount of movement can be reduced by filling the gap with a stiff non-magnetic material.

A.c. motors and generators are also built on laminated cores, so magnetostriction again plays a part in the noise generated, but there are additional sources of magnetic noise. Any eccentricity of the rotor will produce variations in the air gap as the rotor turns. The ideal machine would have the poles uniformly distributed round the rotor and stator, but this is never achieved in practice. A salient pole machine presents the biggest departure from the ideal, but even the slot tooth variation of a distributed winding produces enough variation in the magnetic field to generate magnetic noise. Harmonics in the stator current add to the magnetic noise. Any fluctuations in the electromagnetic force produce vibrations in both the rotor and stator and these also contribute to the noise.

In addition to the purely electromagnetic sources of noise, rotating electrical machines contain the more general noise sources associated with all machinery which we have already discussed. The brushes and brush holders for example can generate noise at discrete frequencies between 2500 and 6500 Hz and also at a frequency given by the product of the rotational speed and the number of commutator segments.

13.8 Noise reduction

In spite of the apparently great diversity in the origins of unwanted noise, there are a relatively small number of techniques of quite general application that may be used to reduce the unwanted noise.

They include the use of a housing to act as an acoustic shield, but, wherever it is proposed to enclose machinery, care must be taken to provide adequate ventilation to prevent overheating. The use of soft materials with high damping is particularly common in gear trains. Where large surface areas are concerned the use of damping coatings may be more appropriate. A rigid construction reduces both the amplitude of any vibration and the irregular wear which would in time lead to noisier operation. Correct alignment during manufacture and installation is always important. Flexible couplings provide a mismatch in the acoustic transmission path where attenuation by reflection can occur, but they are not a substitute for careful assembly. All machinery should be operated as near as possible to its maximum efficiency, and at frequencies well away from any possible frequency of natural vibration.

It is very common for noise generated at one point to travel through a structure or through piping until it reaches a surface which is a good acoustic radiator. In such cases there may be two or more parallel paths, such as the material of the pipe and the water inside the pipe, and treating only one of the paths may not dispose of the whole problem. The condenser of a refrigerator is another example of a component which is a good radiator of noise originally generated elsewhere. Transformers and chokes also radiate much noise from surfaces which do not generate it.

Vibration isolators of all sorts are very useful, but they will not serve their purpose if they are applied without proper thought to the particular application. Nor will they give satisfactory performance if a flanking path is left to bridge them.

Appendix 1

Exponential notation

A mathematical expression for wave motion was derived in Chapter 1, and eqn 1.1 gave it as the partial differential equation

$$\frac{\partial^2 s}{\partial t^2} = \frac{E}{\rho}\frac{\partial^2 s}{\partial x^2} \tag{1.1}$$

or

$$\frac{\partial^2 s}{\partial t^2} = c^2\frac{\partial^2 s}{\partial x^2} \tag{A.1}$$

It was then shown that $s = s_0 \cos 2\pi f(t - x/c)$ is a solution of this equation, and that it represents a travelling wave. It is easy to show that $s = s_0 \sin 2\pi f(t - x/c)$ is an equally good solution which also represents a travelling wave. In fact the two solutions differ only by a phase angle, which is one of the two constants of integration. The other constant is s_0, and if we are prepared to use a complex solution we could equally well write $s = js_0 \sin 2\pi f(t - x/c)$ where $j = \sqrt{-1}$.

Now eqn A.1 is linear, that is, all the differential coefficients in it appear to the first power only. This means that if we have two possible values for s, both of which satisfy eqn A.1, their sum will also be a solution of eqn A.1 giving

$$s = s_0 \cos 2\pi f(t - x/c) + js_0 \sin 2\pi f(t - x/c)$$

It reduces the amount of writing and makes the mathematics easier to read if we write ω instead of $2\pi f$ and k instead of $2\pi f/c$. Then

$$s = s_0 \cos(\omega t - kx) + js_0 \sin(\omega t - kx)$$

$$= s_0 [\cos(\omega t - kx) + j \sin(\omega t - kx)]$$

But since $\cos \theta + j \sin \theta = e^{j\theta}$ we can write

$$s = s_0\, e^{j(\omega t - kx)} \tag{A.2}$$

This is a more general solution of the wave equations 1.1 and A.1. It has all the properties of the simple trigonometric solution and so represents a travelling sinusoidal wave. It has the advantage that it is easier to manipulate mathematically, but the disadvantage that most people find it much harder to visualize its physical significance.

As an example of the use the exponential notation, we will use it to derive the relationship between pressure p and particle velocity u, which has already been found using the trigonometric notation in section 2.3. The pressure p is as before

$$p = -\rho c^2\, \partial s/\partial x$$

$$= -\rho c^2 \times -jks_0 e^{j(\omega t - kx)}$$

$$= j\rho c^2 s_0 k\, e^{j(\omega t - kx)}$$

The velocity u is $u = \partial s/\partial t$

$$= j\omega s_0 \, e^{(\omega t - kx)}$$

Comparing these two expressions, we see that

$$p = \rho c^2 ku/\omega$$
$$= \rho c^2 (2\pi f/c) \, u \, (1/2\pi f)$$
$$= \rho cu$$

As another example, consider the corresponding expression for spherical waves which was found in section 3.3. To derive this using the exponential notation we need the equivalent of the spherical wave equation for pressure eqn 3.9. The exponential form simply has $e^{j(\omega t - kr)}$ instead of $\cos 2\pi(ft - r/\lambda)$, where again $\omega = 2\pi f$ and $k = 2\pi f/c = 2\pi/\lambda$. So eqn 3.9 becomes

$$p = (A/r) \, e^{j(\omega t - kr)}$$

$$\therefore \quad \frac{\partial p}{\partial r} = -\frac{A}{r^2} \, e^{j(\omega t - kr)} - j\frac{kA}{r} \, e^{j(\omega t - kr)}$$

and

$$u = \frac{\partial s}{\partial t} = \frac{-1}{\rho} \int \frac{\partial p}{\partial r} \, dt$$

$$= \int \frac{A}{r^2 \rho} \, e^{j(\omega t - kr)} \, dt + \int \frac{jkA}{r\rho} \, e^{j(\omega t - kr)} \, dt$$

$$= \frac{A}{j\omega r^2 \rho} \, e^{j(\omega t - kr)} + \frac{jkA}{j\omega r\rho} \, e^{j(\omega t - kr)}$$

$$= \frac{kA}{\omega r\rho} \, e^{j(\omega t - kr)} + \frac{jA}{j^2 \omega r^2 \rho} \, e^{j(\omega t - kr)}$$

$$= \left(\frac{kA}{\omega r\rho} - \frac{jA}{\omega^2 r\rho} \right) e^{j(\omega t - kr)}$$

$$= \frac{A}{\omega r^2 \rho} \, (kr - j) \, e^{j(\omega t - kr)} \tag{A.3}$$

The phase difference between p and u is indicated by the j term in the amplitude of $e^{j(\omega t - kr)}$. The phase angle between them is $\theta = \arctan(1/kr) = \arctan(1/(2\pi r/\lambda))$ as already found in Chapter 3.

Complex numbers can be written in various ways, for example

$$x + jy = a \, e^{j\psi}$$

and

$$x - jy = a \, e^{-j\psi}$$

where

$$a = \sqrt{(x^2 + y^2)} \text{ and } \psi = \arctan (y/x)$$

Applying this to eqn A.3 gives

$$u = \frac{A}{\omega r^2 \rho} \sqrt{(k^2 r^2 + 1)} \; e^{-j\theta} \, e^{j(\omega t - kr)}$$

$$= \frac{A}{r} \frac{[(2\pi r/\lambda)^2 + 1]^{\frac{1}{2}}}{2\pi f r \rho} \, e^{j(\omega t - kr - \theta)}$$

$$= \frac{A}{r} \frac{[(2\pi r/\lambda)^2 + 1]^{\frac{1}{2}}}{2\pi (c/\lambda) r \rho} \, e^{j(\omega t - kr - \theta)}$$

$$= \frac{A}{r} \frac{[(2\pi r/\lambda)^2 + 1]^{\frac{1}{2}}}{\rho c (2\pi r/\lambda)} \, e^{j(\omega t - kr - \theta)} \tag{A.4}$$

This is the exponential equivalent of eqn 3.10.

In the first example the exponential notation has no advantage over the trigonometric; in the second example the exponential notation expresses the difference in phase more neatly, if more abstractly; in section 5.3 the succinctness of the exponential notation outweighs all other considerations and makes its use almost essential.

Appendix 2

Sound powers of some sources

Fans

In spite of the many and varied factors which contribute to fan noise, and which might be expected to vary from one model to another, empirical formulae have been found which enable the sound generated by a fan to be predicted with considerable accuracy without referring to the specific details of the design. It is only necessary to know two of the three quantities — volume flow, static head developed, and horse power absorbed. There are three formulae, and whichever two of the quantities are known, one of the formulae will enable the SWL to be calculated. They are

$$SWL = 67 + 10 \log Q + 10 \log p$$
$$SWL = 40 + 10 \log V + 20 \log p$$
$$SWL = 94 + 20 \log Q - 10 \log V$$

where Q = power in kW

p = pressure in N/m^{-2}

V = volume in m^3/s^{-1}

These formulae are consistent with each other if the fan efficiency is 50%. If the efficiency is 80% the discrepancy between them will not exceed 2 dB.

The SWL in each octave band can be found by making the following corrections to the overall SWL. These are approximate only, and can vary from model to model, particularly for axial fans.

Hz	63	125	250	500	1000	2000	4000	8000
Centrifugal	0	−5	−11	−15	−20	−25	−30	−35
Axial	−9	−6	−5	−5	−6	−10	−12	−17

Air diffuser

$$SWL = 13 \log S + 60 \log v - 29$$

where S = neck area in m^2

v = neck velocity in $m\,s^{-1}$

Boiler

$$SWL = 11.5 \log Q + 58 \pm 5 \text{ dB(A)}$$

where Q = output in kW

The sound transmitted through the chimney is given by

$$SPL \text{ in chimney} = 9 \log Q + 78.5$$

Cooling tower

$$SWL = 10 \log Q + 116$$

where Q = fan rating in kW

Electric motors and generators up to about 200 kW

$$SWL = 13 \log Q + 17 \log N + 21$$

where Q = output in kW

N = speed in rev/min

Appendix 3

Acoustic properties of materials

Acoustic velocity and specific acoustic impedance

	c m s^{-1}	ρ kg m^{-3}	ρc kg m^{-2}s^{-1}
Air at 20°C	343	1.21	415
Water	1480	1000	1.48×10^6
Steel	6000	7800	47×10^6
Glass	5300	2640	14×10^6
Concrete	3600	2240	8.1×10^6
Timber	4000	680	2×10^6
Brick	–	1760	
Lead	2160	11340	245×10^6
Breeze block	–	1600	–

Absorption coefficients

	125 Hz	500 Hz	2000 Hz
Brick	0.02	0.02	0.04
Breeze block	0.2	0.6	0.5
Carpet on solid floor	0.1	0.3	0.5
Carpet on timber floor	0.2	0.3	0.5
Concrete	0.01	0.02	0.02
Cork	0.05	0.05	0.1
Curtains	0.05 – 0.1	0.25 – 0.4	0.3 – 0.5
Felt	0.1	0.7	0.8
15 mm fibreboard on solid	0.05	0.15	0.3
15 mm painted fibreboard on solid	0.05	0.1	0.15
15 mm fibreboard on air gap	0.3	0.3	0.3
15 mm painted fibreboard on air gap	0.3	0.15	0.1
Floor tiles	0.03	0.03	0.03
32 oz glass	0.2	0.1	0.05
$\frac{1}{4}$" plate glass	0.1	0.04	0.02
Glazed tiles	0.01	0.01	0.01
25 mm glass or mineral wool on solid	0.2	0.7	0.9
50 mm glass or mineral wool on solid	0.3	0.8	0.75
25 mm glass or mineral wool on air gap	0.4	0.8	0.9
Plaster on solid	0.02	0.02	0.04
Plaster on lath	0.3	0.1	0.04
Plywood on solid	0.05	0.05	0.05
Plywood on air gap	0.3	0.15	0.1
Water	0.01	0.01	0.01
Wood block floor	0.05	0.05	0.1
Wood floor boards	0.15	0.1	0.1
25 mm wood wool on solid	0.1	0.4	0.6
75 mm wood wool on solid	0.2	0.8	0.8
25 mm wood wool on air gap	0.15	0.6	0.6

Equivalent absorption

	125 Hz	500 Hz	2000 Hz
Person in wooden seat	0.17	0.43	0.47
Person in upholstered seat	0.20	0.50	0.55
Wooden seat	0.08	0.16	0.19
Upholstered seat	0.13	0.30	0.34

Appendix 4

Summary of important formulae

Velocity of sound	$c = \sqrt{(E/\rho)}$	(1.6)
Intensity	$I = p^2/\rho c$	(Ch. 2, Ch. 3)
Sound power level	$SWL = 10 \log (W/10^{-12})$ W in watts	
Intensity level	$IL = 10 \log (I/10^{-12})$ I in watts m^{-2}	
Sound pressure level		

$$SPL = 20 \log (p/0.00002) \quad p \text{ in newtons m}^{-2} \quad 2 \times 10^{-5}$$

Sound reduction index	$SRI = -17 + 15 \log \sigma f$	(5.28)
Transmission through ducts		(Ch. 6)

	Attenuation	
	High frequency	Low frequency
Change of area	$10 \log \dfrac{S_1}{S_2}$	$10 \log \dfrac{(1 + S_1/S_2)^2}{4S_1/S_2}$
Branch	$10 \log \dfrac{\Sigma S - S_1}{S_b}$	$10 \log \dfrac{(\Sigma S)^2}{S_1 S_b}$

Free field transmission

$$SPL = SWL - 20 \log r - 11 \tag{7.1}$$

Free field transmission above ground plane

$$SPL = SWL - 20 \log r - 8 \tag{7.2}$$

Attenuation by barrier $= 10 \log (A + B - d) - 10 \log \lambda + 16$ (7.4)

Resonant frequency of membrane $f = \dfrac{c}{2\pi} \sqrt{\left(\dfrac{\rho}{md}\right)}$ (8.6)

$$-60 \sqrt{(1/md)}$$

Attenuation in duct $= 3.5 \, \alpha^{1.4} \, P/S$ (8.10)

Reverberation time $T = 0.161 \, V/a$ (9.6)

or $T = \dfrac{0.161 \, V}{-S \ln (1 - \bar{\alpha})}$ (9.10)

or $T = \dfrac{0.161 \, V}{\Sigma[-S_i \ln (1 - \alpha_i)]}$ (9.11)

Room sound level

$$SPL = SWL + 10 \log \left[\frac{Q}{4\pi r^2} + \frac{4}{R} \right] \qquad (9.16)$$

Transmission between rooms

$$(SPL)_1 - (SPL)_2 = SRI - 10 \log (S_p / S_2 \bar{\alpha}_2) \qquad (9.23)$$

Acoustic power through opening

$$SWL = SPL + 10 \log S - 6 \qquad (9.24)$$

Natural frequency of spring—mass system

$$2\pi f = \sqrt{(s/m)} \qquad (11.4)$$

$$= \sqrt{(g/d)} \qquad (11.6)$$

Addition of decibels

Difference between two levels to be added	Add to larger
0	3
1	3
2	2
3	2
4	2
5	1
6	1
7	1
8	1
9	1
10 or more	0

Wavelengths in air at 20°C

f Hz	63	125	250	500	1000	2000	4000	8000
λ m	5.45	2.74	1.37	0.69	0.34	0.17	0.08	0.04

Appendix 5

Noise criteria

NC values at a given SPL (dB) and frequency

NC	Hz 63	125	250	500	1000	2000	4000	8000
15	47	36	29	22	17	14	12	11
20	51	40	33	26	22	19	17	16
25	54	44	37	31	27	24	22	21
30	57	48	41	35	31	29	28	27
35	60	52	45	40	36	34	33	32
40	64	56	50	45	41	39	38	37
45	67	60	54	49	46	44	43	42
50	71	64	58	54	51	49	48	47
55	74	67	62	58	56	54	53	52
60	77	71	67	63	61	59	58	57
65	80	75	71	68	66	64	63	62

(From Beranek, L. L., *Noise and Vibration Control*, McGraw-Hill, 1957. Used with permission of McGraw-Hill Book Company)

NR values at a given SPL (dB) and frequency

NR	Hz 63	125	250	.500	1000	2000	4000	8000
15	47	36	27	20	15	12	9	7
20	51	40	31	25	20	17	14	12
25	55	44	36	30	25	22	19	18
30	59	48	40	34	29	26	24	23
35	62	52	45	39	34	31	29	28
40	66	56	49	43	39	36	34	33
45	70	61	54	48	44	41	39	38
50	74	65	58	53	49	46	44	43
55	78	69	63	58	54	51	49	48
60	82	74	68	63	59	56	54	53
65	86	78	72	68	64	61	59	58

(From Kosten, G. W., and van Os, G. J., 'Community and reaction criteria from external noises', *NPL Symposium No 12; the Control of Noise*)

Suggested NC or NR values for acceptable background noise

NC or NR	Application
25	Concert halls, broadcasting and recording studios
30	Theatres and cinemas
35	Libraries, museums, court rooms, schools, conference rooms, hospitals, hotels, executive offices, homes
40	Halls, corridors, cloakrooms, restaurants, night clubs, offices, shops
45	Department stores, supermarkets, canteens, general offices
50	Typing pools, offices with business machines
60	Light engineering works
70	Foundries, heavy engineering works

Appendix 6
Log tables

$$10 \log 4\pi = 11 \qquad 10 \log 2\pi = 8$$

Acoustic calculations rarely require more than two-figure accuracy. The following table of logarithms should be adequate for all practical purposes.

	0	1	2	3	4	5	6	7	8	9
10	00	00	01	01	02	02	03	03	03	04
11	04	05	05	05	06	06	06	07	07	08
12	08	08	09	09	09	10	10	10	11	11
13	11	12	12	12	13	13	13	14	14	14
14	15	15	15	16	16	16	16	17	17	17
15	18	18	18	18	19	19	19	20	20	20
16	20	21	21	21	21	22	22	22	23	23
17	23	23	24	24	24	24	25	25	25	25
18	26	26	26	26	26	27	27	27	27	28
19	28	28	28	29	29	29	29	29	30	30
20	30	30	31	31	31	31	31	32	32	32
21	32	32	33	33	33	33	33	34	34	34
22	34	35	35	35	35	35	35	36	36	36
23	36	37	37	37	37	37	37	37	38	38
24	38	38	38	39	39	39	39	39	39	40
25	40	40	40	40	40 ·	41	41	41	41	41
26	42	42	42	42	42	42	42	43	43	43
27	43	43	43	44	44	44	44	44	44	45
28	45	45	45	45	45	45	46	46	46	46
29	46	46	47	47	47	47	47	47	47	48
30	48	48	48	48	48	48	49	49	49	49
31	49	49	49	50	50	50	50	50	50	50
32	51	51	51	51	51	51	51	51	52	52
33	52	52	52	52	52	53	53	53	53	53
34	53	53	53	54	54	54	54	54	54	54
35	54	55	55	55	55	55	55	55	55	56
36	56	56	56	56	56	56	56	56	57	57
37	57	57	57	57	57	57	58	58	58	58
38	58	58	58	58	58	59	59	59	59	59
39	59	59	59	59	60	60	60	60	60	60
40	60	60	60	61	61	61	61	61	61	61
41	61	61	61	62	62	62	62	62	62	62
42	62	62	63	63	63	63	63	63	63	63
43	63	63	64	64	64	64	64	64	64	64
44	64	64	65	65	65	65	65	65	65	65
45	65	65	66	66	66	66	66	66	66	66
46	66	66	66	67	67	67	67	67	67	67
47	67	67	67	67	68	68	68	68	68	68
48	68	68	68	68	68	69	69	69	69	69
49	69	69	69	69	69	69	70	70	70	70
50	70	70	70	70	70	70	70	71	71	71

	0	1	2	3	4	5	6	7	8	9
51	71	71	71	71	71	71	71	71	71	72
52	72	72	72	72	72	72	72	72	72	72
53	72	72	73	73	73	73	73	73	73	73
54	73	73	73	73	74	74	74	74	74	74
55	74	74	74	74	74	74	75	75	75	75
56	75	75	75	75	75	75	75	75	75	76
57	76	76	76	76	76	76	76	76	76	76
58	76	76	76	77	77	77	77	77	77	77
59	77	77	77	77	77	77	78	78	78	78
60	78	78	78	78	78	78	78	78	78	78
61	79	79	79	79	79	79	79	79	79	79
62	79	79	79	79	80	80	80	80	80	80
63	80	80	80	80	80	80	80	80	80	81
64	81	81	81	81	81	81	81	81	81	81
65	81	81	81	81	82	82	82	82	82	82
66	82	82	82	82	82	82	82	82	82	83
67	83	83	83	83	83	83	83	83	83	83
68	83	83	83	83	84	84	84	84	84	84
69	84	84	84	84	84	84	84	84	84	84
70	85	85	85	85	85	85	85	85	85	85
71	85	85	85	85	85	85	85	86	86	86
72	86	86	86	86	86	86	86	86	86	86
73	86	86	86	87	87	87	87	87	87	87
74	87	87	87	87	87	87	87	87	87	87
75	88	88	88	88	88	88	88	88	88	88
76	88	88	88	88	88	88	88	88	89	89
77	89	89	89	89	89	89	89	89	89	89
78	89	89	89	89	89	89	90	90	90	90
79	90	90	90	90	90	90	90	90	90	90
80	90	90	90	90	91	91	91	91	91	91
81	91	91	91	91	91	91	91	91	91	91
82	91	91	91	92	92	92	92	92	92	92
83	92	92	92	92	92	92	92	92	92	92
84	92	92	93	93	93	93	93	93	93	93
85	93	93	93	93	93	93	93	93	93	93
86	93	94	94	94	94	94	94	94	94	94
87	94	94	94	94	94	94	94	94	94	94
88	94	95	95	95	95	95	95	95	95	95
89	95	95	95	95	95	95	95	95	95	95
90	95	95	96	96	96	96	96	96	96	96
91	96	96	96	96	96	96	96	96	96	96
92	96	96	96	97	97	97	97	97	97	97
93	97	97	97	97	97	97	97	97	97	97
94	97	97	97	97	98	98	98	98	98	98
95	98	98	98	98	98	98	98	98	98	98
96	98	98	98	98	98	98	99	99	99	99
97	99	99	99	99	99	99	99	99	99	99
98	99	99	99	99	99	99	99	99	99	99
99	99	99	99	99	99	99	99	99	99	99

Appendix 7

Basic trigonometric formulae

$$\cos A \cos B = \tfrac{1}{2} \cos (A + B) + \tfrac{1}{2} \cos (A - B)$$

$$\cos^2 A = \tfrac{1}{2}(1 + \cos 2A)$$

$$a \sin \theta + b \cos \theta = \sqrt{(a^2 + b^2)}(\sin \theta \sin \phi + \cos \theta \cos \phi)$$

$$= \sqrt{(a^2 + b^2)} \cos (\theta + \phi)$$

where

$$\sin \phi = a/\sqrt{(a^2 + b^2)}$$

$$\cos \phi = b/\sqrt{(a^2 + b^2)}$$

$$\tan \phi = a/b$$

$$\cos A + \cos B = 2 \cos \tfrac{1}{2}(A + B) \cos \tfrac{1}{2}(A - B)$$

Bibliography

Anthrop, D. F., *Noise Pollution*, Lexington Books, 1973
  Descriptive, with emphasis on American legal requirements.
Beranek, L. L., ed, *Noise and Vibration Control*, McGraw-Hill, Maidenhead, 1957
  Detailed mathematical treatment directed towards engineering applications.
Blitz, J., *Elements of Acoustics,* Butterworth, London, 1964
  Theoretical, no details of application.
Bradbury, C. H., *Engine Noise: Analysis and Control*, Temple Press Books, London, 1963
  Excellent introductory treatment limited to scope of subject.
Brüel, P. V., *Sound Insulation and Room Acoustics,* Chapman and Hall, London, 1951
  Much detailed information.
Constable, J. E. R. and Constable, K. M., *The Principles and Practice of Sound Insulation*, Pitman, London, 1949
  Excellent detailed descriptions with much useful data.
Cremer, L., Heckel, M., Under, E. E., *Structure-bourne Sound*, Springer-Verlag, 1973
  Mathematical theory of vibration.
Ford, R. D., *Introduction to Acoustics*, Applied Science Publishers, London, 1970
  Theoretical development with no emphasis on applications.
Furrer, W., *Room and Building Acoustics and Noise Abatement*, Butterworths, London, 1964
  Detailed practical information.
Geiger, P. H., *Noise Reduction Manual*, Engineering Research Institute, University of Michigan, 1955
  Details practical manual with no theoretical derivation.
Harris, C. M., ed., *Handbook of Noise Control*, McGraw-Hill, Maidenhead, 1957
  Detailed information, both theoretical and practical, on almost every application.
Hines, W. A., *Noise Control in Industry*, Business Publications Limited, 1966
  Purely descriptive.
Kerse, C. S., *The Law Relating to Noise*, Oyez Publishing, London, 1975
  Fully covers English law.
King, A. J., *The Measurement and Suppression of Noise*, Chapman and Hall, London, 1965
  Excellent guide to practical application with some indication of theory.
Kinsler, L. E. and Frey, A. R., *Fundamentals of Acoustics*, John Wiley, 2nd edition, New York, 1962
  Good mathematical introduction. A shorter and more modern treatment than Rayleigh.
Lord, P. and Thomas, F. L., ed., *Noise Measurement and Control*, Heywood, London, 1963
  Descriptive but very detailed.

Meyer, E. and Neumann E. G., *Physical and Applied Acoustics*, Academic Press, London, 1972
  Advanced mathematical treatment.
Parkin, P. H. and Humphreys, H. R., *Acoustics, Noise and Buildings*, Faber, London, 1958
  Practical, with little theoretical background.
Rayleigh, Baron (Strutt, J. W.), *The Theory of Sound*, 2 volumes, Macmillan, 2nd edition, London, 1926
  The classic and classical foundation, still worth reading.
Richardson, E. G., ed., *Technical Aspects of Sound*, 2 volumes, Elsevier, Amsterdam, 1953
  Detailed and exhaustive.
Richardson, E. G., *Sound*, Edward Arnold, 5th edition, London, 1953
  The physics of sound.
Sharland, I., *Woods Practical Guide to Noise Control*, Woods of Colchester, 1972
  Practical and elementary, no theory.
Warring, R. H., ed., *Handbook of Noise and Vibration Control*, Trade and Technical Press, Morden, 1973
  Detailed, with more emphasis on practical applications than theory.
Watson, F. R., *Acoustics of Buildings*, John Wiley, 3rd edition, Chichester, 1941
  A wealth of information limited to scope of subject.
Wood, A. B., *A Textbook of Sound*, G. Bell, 3rd edition, London 1955
  The physics of sound.
Yerges, L. F., *Sound, Noise and Vibration Control*, Van Nostrand Reinhold, London, 1969
  Very detailed, with much practical performance data.

Index